D068331?

Physics
for Cambridge IGCSE®

Revision Guide

Sarah Lloyd

OXFORD
UNIVERSITY PRESS

OXFORD
UNIVERSITY PRESS

Great Clarendon Street, Oxford OX2 6DP

Oxford University Press is a department of the University of Oxford.
It furthers the University's objective of excellence in research,
scholarship, and education by publishing worldwide in

Oxford   New York

Auckland   Cape Town   Dar es Salaam   Hong Kong   Karachi
Kuala Lumpur   Madrid   Melbourne   Mexico City   Nairobi
New Delhi   Shanghai   Taipei   Toronto

With offices in

Argentina   Austria   Brazil   Chile   Czech Republic   France   Greece
Guatemala   Hungary   Italy   Japan   Poland   Portugal   Singapore
South Korea   Switzerland   Thailand   Turkey   Ukraine   Vietnam

Oxford is a registered trade mark of Oxford University Press
in the UK and in certain other countries

British Library Cataloguing in Publication Data

Data available

ISBN: 978-0-19-915436-4
10 9 8 7

Printed in Great Britain by Bell & Bain Ltd, Glasgow

® IGCSE is the registered trademark of Cambridge International Examinations.

Past paper examination material reproduced by permission of Cambridge International Examinations.

Cambridge International Examinations bears no responsibility for the example answers to questions taken
from its past question papers which are contained in this publication.

Acknowledgements

The publishers would like to thank the following for permission to reproduce photographs:

Cover photo: Don Farrall/Getty Images

Paper used in the production of this book is a natural, recyclable product made from wood grown in
sustainable forests. The manufacturing process conforms to the environmental
regulations of the country of origin.

Contents

Answers provided on our website: www.oxfordsecondary.co.uk/physicsrg

How to use this revision guide

This book is designed to be used with *Complete Physics for IGCSE*. It offers brief notes and simplified explanations, along with practice questions, to help you understand the physics principles required for the Cambridge IGCSE syllabus. The notes, examples, summary questions and examination questions are divided into sections that relate to the syllabus areas.

The examination questions that accompany each subsection (e.g. transfer of thermal energy, which is part of the thermal physics section) will allow you to test yourself at regular intervals. There are also questions at the end of each section so that you can test your knowledge and understanding of the whole topic. In this way you can revise topic by topic until you have covered the entire syllabus. Answers to questions are available at www.OxfordSecondary.co.uk

Tips for effective revision

Active Revision for Physics

Making revision notes

Don't just write out your notes. Try to make them as brief as possible, just picking out the essential points. This is quite challenging! You could try writing your revision notes onto cue cards. They are then more portable to read on the bus/tram to school or on the way to see your friends.

Drawing revision mind maps (spider diagrams)

This is a good way to visually summarise information. You can link ideas, which will help your understanding of the topic. Simplifying difficult concepts into diagrams will help you to reduce the information that you have to learn for the exam. Drawing out a mind map on a large sheet of paper will allow you to put in more diagrams and highlight the important points in colour. Below is an example of a simple mind map on radioactivity.

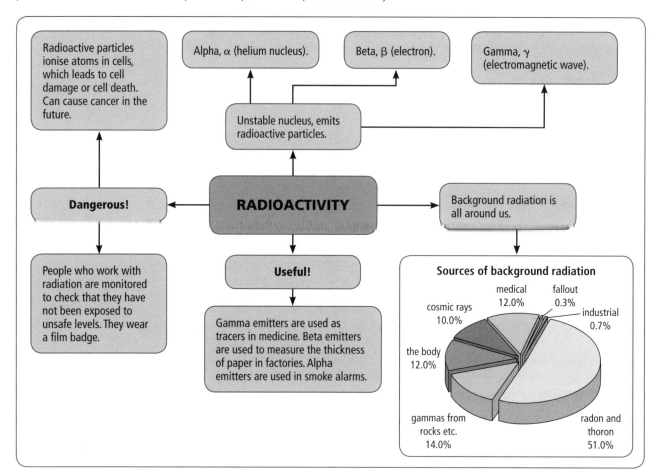

Writing your own test

Write down questions as you go through this Revision Guide. When you reach the end of the topic you can use the questions to test your knowledge and understanding and check your progress.

Getting someone else to test you

You could give one of your classmates your revision notes and ask them to test you. Or you could get them to ask you about parts of the syllabus they don't understand. This is a good test of your understanding of that topic!

Using your syllabus

You can download a copy of the syllabus from the CIE website. Make check lists for revision using the syllabus to indicate when you have made revision notes, practised exam questions etc. Don't forget to include a column to tick when you feel you have understood a topic.

Doing past exam questions

There are lots of examination questions throughout this book for you to try. Try revising a small part of a topic, say for about 30 minutes and then test yourself on the examination style questions provided here. When you feel ready, full past papers are available on the CIE website

Making revision posters

If you make your revision notes into posters and put them in places where you will see them often, you will read them without even realising! This will help to keep topics fresh in your mind. Your family and friends will see what you have been revising and might talk to you about it, which will help the information in stick your mind.

Essential quantities and units

Quantity	Symbol	Unit
Time	t	second (s)
Force	F	newton (N)
Weight	W	newton (N)
Velocity	v	metres per second (m/s)
Speed	s	metres per second (m/s)
Distance	d or s	metre (m)
Acceleration	a	metres per second squared (m/s^2)
Mass	m	kilogram (kg)
Moment	M	newton metre (Nm)
Energy	E	joule (J)
Work done	W	joule (J)
Power	P	watt (W)
Current	I	ampere (A)
Potential difference (voltage)	V	volt (V)
Resistance	R	ohm (Ω)
Charge	Q	coulomb (C)
Frequency	f	hertz (Hz)
Pressure	p	newton per metre squared (N/m^2)
Temperature	T	degrees Celsius (°C)
Density	ρ or d	g/cm^3 or kg/m^3
Wavelength	λ	metre (m)
Specific heat capacity	c	joules per kilogram degree Celsius J/(kg °C)
Specific latent heat	l	joules per kilogram (J/kg)

Formulae and magic triangles

These magic triangles are a useful way of remembering how to use an equation, but are no substitute for remembering the equation itself.

To use a magic triangle, cover the quantity you are trying to find. The relationship left behind shows you how to calculate it. For example, cover up speed, u in the first triangle and you are left with $\frac{d}{t}$.

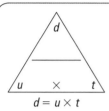

$$d = u \times t$$

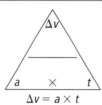

$$\Delta v = a \times t$$

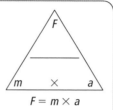

$$F = m \times a$$

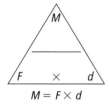

$$M = F \times d$$

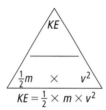

$$KE = \frac{1}{2} \times m \times v^2$$

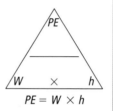

$$PE = W \times h$$

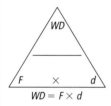

$$WD = F \times d$$

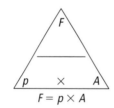

$$F = p \times A$$

$$E = p \times t$$

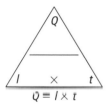

$$Q = I \times t$$

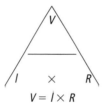

$$V = I \times R$$

$$P = I \times V$$

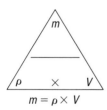

$$m = \rho \times V$$

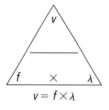

$$v = f \times \lambda$$

$$p = \rho g h$$

$$E = mc\Delta T$$

$$E = \Delta m L$$

1 General physics

1.1 Length, volume and time

KEY IDEAS

✓ Length is measured with a ruler, tape measure, vernier callipers or micrometer screw gauge
✓ Volume may be measured with a measuring cylinder
✓ Time is measured with a stop clock or stop watch

Length measurements can be made more **precise** by using an instrument with a vernier scale such as vernier callipers or a micrometer screw gauge. A ruler measures to the nearest millimetre, vernier callipers to 1/10 mm and a micrometer 1/100 mm. Length measurements can be made more **accurate** by measuring multiples, such as the thickness of 500 sheets of paper, then divide by 500 to get the thickness of one sheet. Check the **reliability** of your measurements by repeating them. If the repeated results are similar, they are reliable. When measuring the length, l of a pendulum as in the figure (right), you need to measure to the centre of gravity. One way of doing this is to take two measurements and average them: from the fixed end of the string to the beginning of the bob, l_1 and from the fixed end of the string to the far end of the bob l_2.

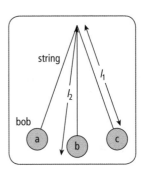

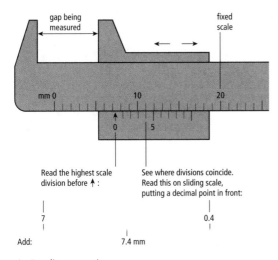

▲ Reading a vernier

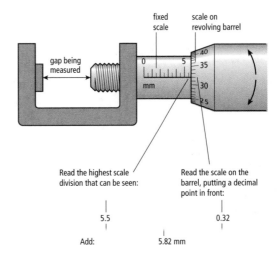

▲ Reading a micrometer

The volume of a regular solid can be found by measuring its dimensions. For example, recording the length, width and height of a cuboid (box) and multiplying these measurements together. The volume of an irregularly shaped object can also be measured using a measuring cylinder with a eureka can. Place the object in the eureka can and use the measuring cylinder to measure how much water is displaced. The volume of water displaced is equal to the volume of the object.

A digital stop watch measures time to a **precision** of 0.01 s. This is far more precise than human error will allow, which is about 0.2 s. If possible, time over as long a period as possible. When timing the time period for a simple pendulum, for example, it is more **accurate** to time 100 swings with a stop watch and then divide by the number of swings. This reduces the error due to human reaction time. Repeat time measurements to check for **reliability** of data.

▲ Eureka can and measuring cylinder

1.2 Speed, velocity and acceleration

KEY IDEAS

✓ Speed $= \dfrac{\text{distance}}{\text{time}}$

✓ Acceleration $= \dfrac{\text{change in velocity}}{\text{time}}$

✓ Velocity and acceleration can have both positive and negative values

$$\text{Speed (m/s)} = \frac{\text{distance (m)}}{\text{time (s)}}$$

$$u = \frac{d}{t}$$

Worked examples

1. A car travels at a speed of 20 m/s for 30 s. How far does it travel in this time?

2. A cyclist travels 1000 m in 3 minutes. What is his speed?

3. A girl walks 3 km at 1.5 m/s. How long does her journey take?

Answers

1. $d = u \times t$
$= 20 \times 30$
$= 600$ m

2. $u = d/t$
$= 1000 \div (3 \times 60)$
$= 1000 \div 180$
$= 5.56$ m/s

3. $t = d/u$
$= (3 \times 1000) \div 1.5$
$= 3000 \div 1.5$
$= 2000$ s (33 min 20 s)

Note: in example 2, time in minutes must be converted to **seconds**. In example 3, distance in kilometres must be converted to **metres**.

Distance–time graphs

A journey can be represented on a graph by plotting the distance travelled on the *y*-axis and the time taken on the *x*-axis. The shape of the graph describes the journey.

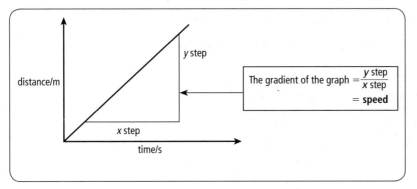

The gradient of the graph $= \dfrac{y \text{ step}}{x \text{ step}}$
$= \textbf{speed}$

▲ Distance–time graph

Examples of distance–time graphs

1. In a crash test, a car travels at steady speed and then stops suddenly as it hits a wall.

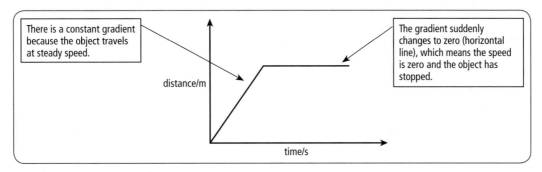

There is a constant gradient because the object travels at steady speed.

The gradient suddenly changes to zero (horizontal line), which means the speed is zero and the object has stopped.

2. A runner sets off in a race, increasing her speed until she reaches her maximum speed.

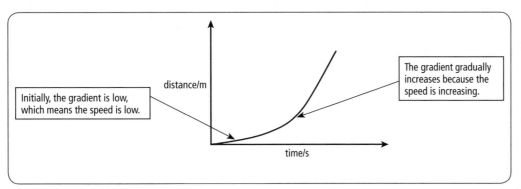

Initially, the gradient is low, which means the speed is low.

The gradient gradually increases because the speed is increasing.

Velocity and acceleration

Velocity is a **vector** quantity; **it is equal to speed in a particular direction**. Speed is a measurement of how fast an object is moving.

Acceleration is also a vector quantity. It is equal to the **change in velocity per second**.

When an object is slowing down, if its velocity is **decreasing**, the acceleration is **negative**. We say that it is **decelerating**. See section 1.5 (scalars and vectors).

Acceleration (m/s^2) = change in velocity (m/s) ÷ time (s)

$$a = \frac{v - u}{t}$$

$$a = \frac{\Delta v}{t}$$

initial velocity = u, final velocity = v,
time taken for the change in velocity = t,
change in velocity = $v - u = \Delta v$

Worked examples

1. A car increases its velocity from 10 m/s to 20 m/s in 5 s. What is its acceleration?

2. A runner has an acceleration of 10 m/s^2. How long does it take him to reach a speed of 5 m/s from rest? (Note 'rest' means zero velocity.)

3. A train accelerates at 9 m/s^2 for 5 s. If its initial velocity is 5 m/s, what is its final velocity?

Answers

1. $a = \Delta v \div t$
$\quad = (20 - 10) \div 5$
$\quad = 2$ m/s^2

2. $t = \Delta v \div a$
$\quad = 5 \div 10$
$\quad = 0.5$ s

3. $\Delta v = a \times t$
$\quad = 9 \times 5$
$\quad = 45$ m/s
$v = u + \Delta v$
$\quad = 5 + 45$
$\quad = 50$ m/s

Speed–time graphs

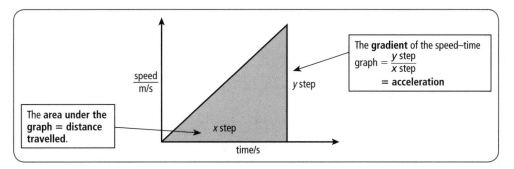

▲ Speed–time graph.

The area under the graph is a calculation involving the units on the two axes. It is not a physical area.

Examples of speed–time graphs

1. A car accelerating until it reaches its maximum speed.

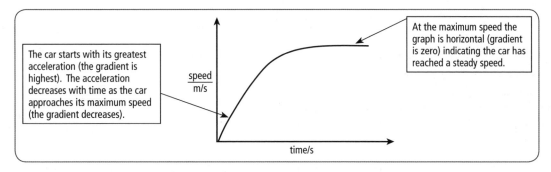

2. A runner who accelerates with constant acceleration to his maximum speed and then decelerates steadily to a stop at the end of the race.

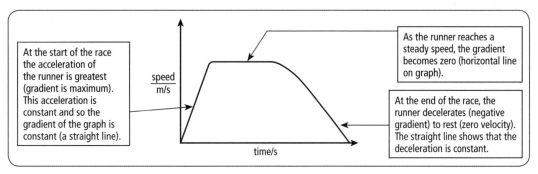

3. A skydiver from the time she jumps from a helicopter until the moment she reaches the ground.

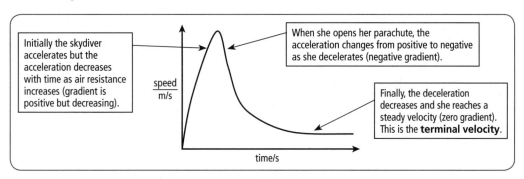

Examination style questions

1.

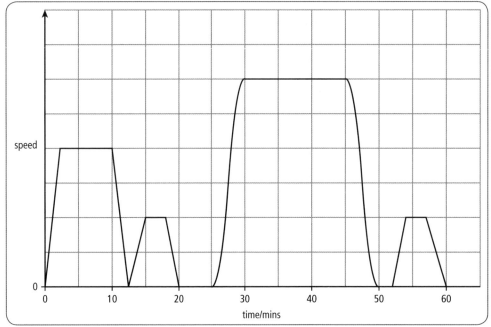

The graph above is for a 60 minute car journey.
a.   Between which times is the car speed at its highest?
b.   Calculate the total time for which the car is stopped.
c.   State without calculation how the graph could be used
    i)   to find the distance travelled in the first 12½ minutes.
    ii)   to find the average speed for the journey.

Graph adapted from CIE 0625 June '05 Paper 2 Q2

2. A stone falls from the top of a building and hits the ground at a speed of 32 m/s.
The air resistance force on the stone is very small and may be neglected.
a.   Calculate the time of fall.
b.   Copy and draw the speed–time graph for the falling stone.

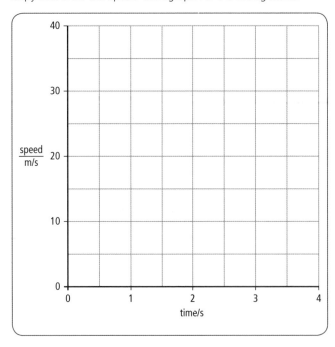

CIE 0625 November '06 Paper 3 Q1a

3. The figure below shows the speed–time graph for a journey travelled by a tractor.

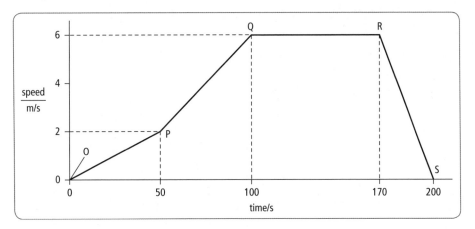

a.   Use the graph to describe the motion of the tractor during the sections OP, PQ, QR and RS.
b.   Which two points on the graph show when the tractor is stationary?
c.   State the greatest speed reached by the tractor.
d.   For how long was the tractor travelling at constant speed?
e.   State how the graph may be used to find the total distance travelled during the 200 s journey. Do **not** attempt a calculation.

CIE 0625 November '06 Paper 2 Q3

4. Palm trees are growing every 25 m alongside the highway in a holiday resort.

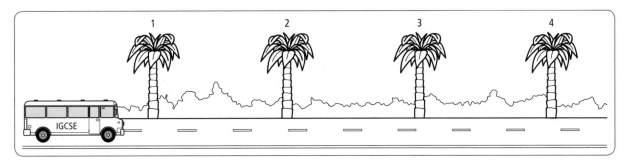

The IGCSE school bus drives along the highway.
a.   It takes 2 s for the bus to travel between palm tree 1 and palm tree 2.
     Calculate the average speed of the bus between tree 1 and tree 2.
b.   It takes more than 2s for the bus to travel from tree 2 to tree 3.
     State what this information indicates about the speed of the bus.
c.   The speed of the bus continues to do what you have said in (b). State how the time taken to go from tree 3 to tree 4 compares with the time in (b).

CIE 0625 November '05 Paper 2 Q2

5. In a training session, a racing cyclist's journey is in three stages.

Stage 1   He accelerates uniformly from rest to 12 m/s in 20 s.

Stage 2   He cycles at 12 m/s for a distance of 4800 m.

Stage 3   He decelerates uniformly to rest.

The whole journey takes 500 s.

a.   Calculate the time taken for stage 2.

b.   Copy the grid below and draw a speed–time graph of the cyclist's ride.

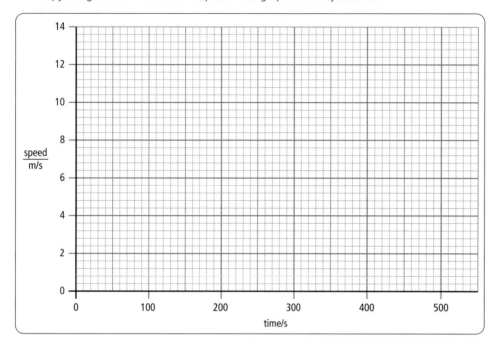

c.   Show that the total distance travelled by the cyclist is 5400 m.

d.   Calculate the average speed of the cyclist.

CIE 0625 June '07 Paper 2 Q2

1.3 Mass and weight

KEY IDEAS

✓ Mass is a quantity related to the inertia of an object, measured in kg
✓ Weight is the force, in N, on a mass due to a gravitational field
✓ On Earth, the gravitational field strength is 10 N/kg

Mass is the amount of **matter** that makes up an object. It is measured in kilograms (**kg**). All masses have a quality called "inertia", the tendency to keep moving if already moving and stay still if already still.

For example, a car in a crash test: when the car hits the wall, it decelerates to rest in a short time. The crash test dummy has inertia due to its mass and so it keeps moving forwards at the same speed as before the car hit the wall. Although it looks as though the dummy has been thrown forward, there is no net forward force on it.

Weight is the force on a mass due to gravity. It is measured in newtons (**N**).

weight (N) = mass (kg) × gravitational field strength (N/kg)
$W \quad = \quad m \quad \times \quad g$

On Earth, the gravitational field strength, $g = 10$ **N/kg**. This is also called the acceleration due to gravity or the acceleration of freefall and has an alternative unit of m/s^2.

Examination style question

Some IGCSE students were asked to write statements about mass and weight.

Their statements are printed below. Choose the **two** correct statements.

Mass and weight are the same thing.

Mass is measured in kilograms.

Weight is a type of force.

Weight is the acceleration caused by gravity.

CIE 0625 November '06 Paper 2 Q2

1.4 Density

Density is a quantity related to how closely packed the particles in a material are as well as how much the particles weigh.

$$\text{density (g/cm}^3) = \frac{\text{mass (g)}}{\text{volume (cm}^3)}$$
$$\rho = \frac{m}{V}$$

A simple method of measuring the density of an object (if its density is greater than that of water):

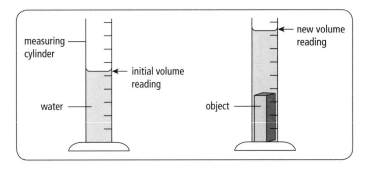

- Find the mass of the object using a balance.
- Approximately half fill a measuring cylinder with water. Read the volume of water from the measuring cylinder scale.
- Place the object in the measuring cylinder and take the new volume reading.
- Calculate (new volume reading – initial volume reading) to find the volume of the object.
- Calculate the density by dividing the mass by the volume.

Worked examples

1. The mass of a stone is found on a top pan balance. It has a mass of 120.02 g. A measuring cylinder is filled to a volume of 60 cm³ with water. When the stone is placed in the measuring cylinder, the new water level is 95 cm³. Find the density of the stone.

2. An object of known density of 2.7 g/cm³ is placed into a measuring cylinder of water. The level of the water rises from 45 cm³ to 72 cm³. What is the mass of the object?

3. A metal block of density 3.2 g/cm³ and mass 90 g is placed in a measuring cylinder containing 65 cm³ of water. What is the new water level?

Answers

1. Volume = 95 − 60 = 35 cm³
$\rho = \frac{m}{V} = \frac{120.02}{35} = 3.4$ g/cm³

2. Volume = 72 − 45 = 27 cm³
$m = V \times \rho = 27 \times 2.7 = 73$ g

3. $V = \frac{m}{\rho} = \frac{90}{3.2} = 28$ cm³
New water level = 65 + 28 = 93 cm³

Examination style question

1. A student is given a spring balance with a Newton scale. She is told that the acceleration due to gravity is 10 m/s^2.
Describe how she could find the mass of a toy car.
Describe how she could go on to find the average density of the toy car.

Adapted from CIE 0625 November '05 Paper 3 Q1b

2. A student used a suitable measuring cylinder and a spring balance to find the density of a sample of a stone.
   i)   Describe how the measuring cylinder is used, and state the readings that are taken.
   ii) Describe how the spring balance is used, and state the reading that is taken.
   iii) Write down an equation from which the density of the stone is calculated.
   iv) The student then wishes to find the density of cork. Suggest how the apparatus and the method would need to be changed.

CIE 0625 November '06 Paper 3 Q1b

3. Fig. (a) shows a measuring cylinder, containing some water, on a balance.
Fig. (b) shows the same arrangement with a stone added to the water.

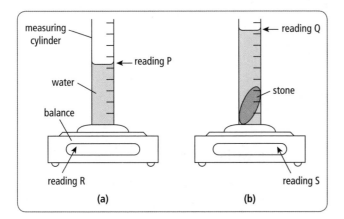

a. Which two readings should be subtracted to give the volume of the stone?
b. Which two readings should be subtracted to give the mass of the stone?
c. In a certain experiment,

$$\text{mass of stone} = 57.5 \text{ g},$$
$$\text{volume of stone} = 25 \text{ cm}^3.$$

   i) Write down the equation linking density, mass and volume.
   ii) Calculate the density of the stone.

CIE 0625 June '06 Paper 2 Q3

Practical question

An IGCSE student is determining the density of a metal alloy.

The student is provided with several metal rods, as shown on the right.

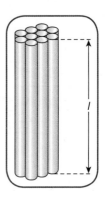

1. Measure with a ruler the length, l, of one of the rods.

2. The student measured the diameter of one of the rods with a ruler and found it to be 0.6 cm. Calculate the cross-sectional area, A, of the rod.

3. Use this value to calculate the volume, V, of one rod and hence the whole bundle.

The student used a balance to find the mass of the bundle and found it to be 59.1 g. Calculate the density of the metal alloy.

1.5 Forces

KEY IDEAS

✓ Forces are vector quantities that can change the shape of an object, accelerate it or change its direction
✓ Moment of a force about a pivot = force × perpendicular distance from the pivot
✓ An object is in equilibrium if there is zero net force on it and zero net moment
✓ An object will not fall over if the line of action of its weight acts through its base; this depends on the position of the centre of mass
✓ The resultant of two forces can be found by drawing a scale diagram

Forces can produce a **change in size or shape of an object**. For example, loading a spring will increase its length.

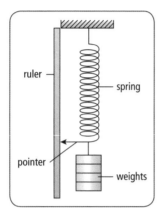

An experiment to find how the extension of a spring varies with the force applied

- Measure the original position of the spring using the pointer and ruler, to the nearest mm.
- Add a 100 g mass hanger and measure the new position.
- Repeat 6 times, adding a 100 g mass each time.
- Calculate the extension by subtracting the original position from each subsequent position reading.
- Plot a graph of extension against force, where force = mass × 10 N/kg.

Graph of results

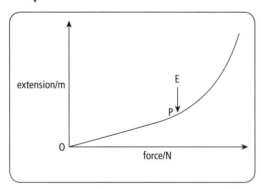

Conclusion

From **0** to **P**, extension is directly proportional to the force applied. **Beyond P**, the extensions are larger for the same increase in force.

At **E**, the elastic limit is reached. Beyond this point, the spring will not return to its original length when the force is removed.

Hooke's Law

If a material obeys Hooke's Law, the extension is directly proportional to the applied force, provided that the elastic limit is not exceeded.

$$F = kx$$

Where F = applied force (N), k = force constant for object under test (N/m), x = extension (m)

Force, mass and acceleration

A force can **accelerate** an object. The larger the force on the object, the greater the acceleration if the mass stays constant. The larger the mass of the object, the smaller the acceleration if the force stays constant.

$$\text{Force (N)} = \text{mass (kg)} \times \text{acceleration (m/s}^2)$$

$$F \quad = \quad m \quad \times \quad a$$

Worked examples

1. A force of 10 N acts on an object of mass 5 kg. What is the acceleration of the object?

2. A force of 15 N causes an object to accelerate at 2 m/s². What is its mass?

3. A mass of 3 kg has a deceleration of 5 m/s². What force acts on it?

Answers

1. $a = F \div m$
$= \frac{10}{5}$
$= 2$ m/s²

2. $m = F \div a$
$= \frac{15}{2}$
$= 7.5$ kg

3. $F = m \times a$
$= 3 \times -5$
$= -15$ N

Note: in example 3, the object is decelerating so acceleration is negative.

Resultant force

The **resultant** force acting on an object is the **net** or **overall** force when the **size** and **direction** of all the forces acting are taken into account. **Force** is a **vector** quantity.

Examples

1.

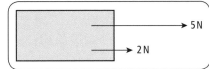

Resultant force = 5 N + 2 N = 7 N

The forces are **added** together because they act in the **same** direction.

2.

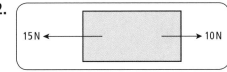

Resultant force = 15 N − 10 N
= 5 N (to the left)

The forces are **subtracted** because they act in **opposite** directions.

A force can cause an object to **change direction**. The object will move in a circle if the force acts perpendicular (at a right angle) to the direction that the object is travelling.

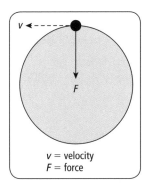

v = velocity
F = force

The force acts at **right angles** to the direction that the object is moving. The force does not do any work on the object because the object does not move in the direction of the force. The force constantly changes the direction of the object which means its **velocity changes** but its **speed stays constant**. The object accelerates towards the centre of the circle.

The force which causes an object to move in a circle is called the **centripetal force**. The centripetal force increases if:

- the mass of the object increases
- the speed of the object increases
- the radius of the circle decreases

If the force which is providing the centripetal acceleration is suddenly removed, the object will move on a tangent to the original circle.

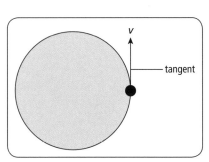

Examination style questions

1.

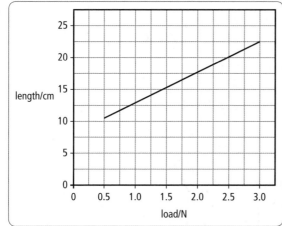

The graph above was obtained by a student who loaded a spring from 0.5 N to 3.0 N and measured its length.

a.   Find the length of the spring when a load of 2.0 N is applied

b.   Find the load required to stretch the spring to a length of 18 cm

c.   Find the extension of the spring when it is stretched by a load of 1.5 N

Adapted from CIE 0625 November '05 Paper 2 Q1

2. A mass of 5.0 kg accelerates at 3 m/s² in a straight line.

a.   State why acceleration is described as a vector quantity.

b.   Calculate the force required to accelerate the mass.

Adapted from CIE 0625 June '05 Paper 3 Q3

3. The length of a spring is measured when various loads from 1.0N to 6.0N are hanging from it. The graph below gives the results.

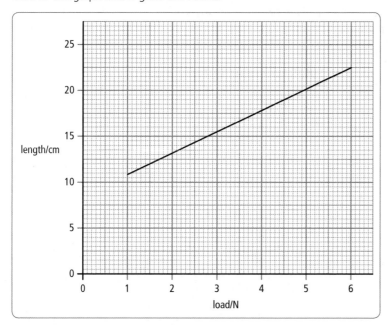

Use the graph to find

a.   the length of the spring with no load attached,

b.   the length of the spring with 4.5N attached,

c.   the extension caused by a 4.5N load.

CIE 0625 November '05 Paper 2 Q1

4. In an experiment, forces are applied to a spring as shown in (a). The results of this experiment are shown in (b).

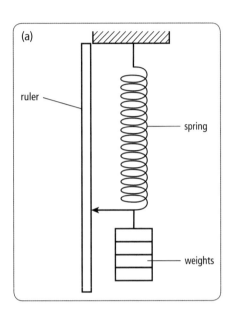

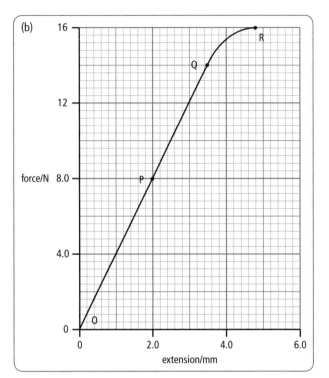

a. What is the name given to the point marked Q in (b)?
b. For the part OP of the graph, the spring obeys Hooke's Law.
   State what this means.
c. The spring is stretched until the force and extension are shown by the point R on the graph. Compare how the spring stretches, as shown by the part of the graph OQ, with that shown by QR.
d. The part OP of the graph shows the spring stretching according to the expression
$$F = kx.$$
   Use values from the graph to calculate the value of k.

CIE 0625 November '06 Paper 3 Q2

5. The points plotted on the grid below were obtained from a spring-stretching experiment.

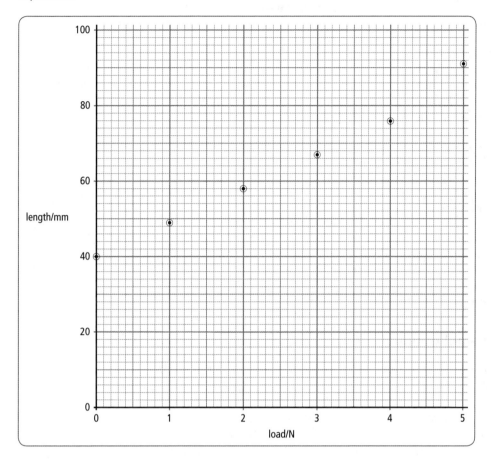

a. Using a straight edge, draw a straight line through the first 5 points. Extend your line to the edge of the grid.
b. Suggest a reason why the sixth point does not lie on the line you have drawn.
c. Calculate the extension caused by the 3N load.
d. A small object is hung on the unloaded spring, and the length of the spring becomes 62 mm.
   Use the graph to find the weight of the object.

CIE 0625 November '06 Paper 2 Q9

Practical question

An IGCSE class is investigating the effect of a load on a metre rule attached to a spring. The apparatus is shown in the diagram below.

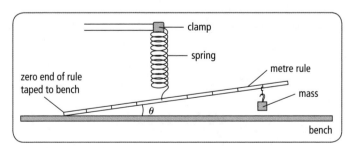

The zero end of the metre rule is taped to the bench to stop it slipping. The spring is attached to the rule at the 40.0 cm mark and the masses are attached at the 90.0 cm mark. The masses are added 10 g at a time and the angle, θ, between the bench and the rule measured with a protractor.

One student's results are shown below

m	θ
0	29
10	28
20	26
30	25
40	22
50	19

1. Complete the column headings.

2. One student suggests that m and θ should be directly proportional to each other. Plot a graph of θ (y-axis) against m (x-axis). Using your graph show whether this prediction is correct. State your reason.

Turning effect and equilibrium

The turning effect or **moment** of a force about a pivot is equal to the force multiplied by its perpendicular distance from the pivot.

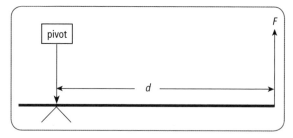

Moment (Nm) = Force (N) × distance (m)

$$M \quad = \quad F \quad \times \quad d$$

If an object is in equilibrium there is no resultant turning effect and no resultant force.

Worked example

1. A force of 2.0 N acts at distance of 3.0 m from a pivot. Find the moment of the force.

2. A force of 5.0 N provides a moment of 15 Nm about a pivot. What is the distance of the force from the pivot?

3. A force provides a moment of 20 Nm about a pivot at a distance of 2.0 m. What is the size of the force?

Answers

1. $M = F \times d$
   $= 2.0 \times 3.0$
   $= 6.0 \text{ Nm}$

2. $d = \dfrac{M}{F}$
   $= \dfrac{15}{5.0}$
   $= 3.0 \text{ m}$

3. $F = \dfrac{M}{d}$
   $= \dfrac{20}{2.0}$
   $= 10 \text{ N}$

Examples of objects in equilibrium

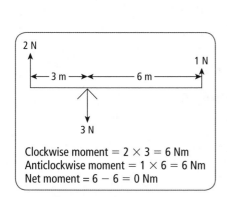

Clockwise moment $= 2 \times 3 = 6$ Nm
Anticlockwise moment $= 1 \times 6 = 6$ Nm
Net moment $= 6 - 6 = 0$ Nm

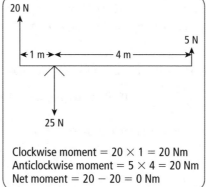

Clockwise moment $= 20 \times 1 = 20$ Nm
Anticlockwise moment $= 5 \times 4 = 20$ Nm
Net moment $= 20 - 20 = 0$ Nm

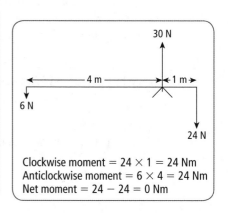

Clockwise moment $= 24 \times 1 = 24$ Nm
Anticlockwise moment $= 6 \times 4 = 24$ Nm
Net moment $= 24 - 24 = 0$ Nm

Taking moments about the pivot in each case.

An experiment to show that there is no net moment on an object in equilibrium

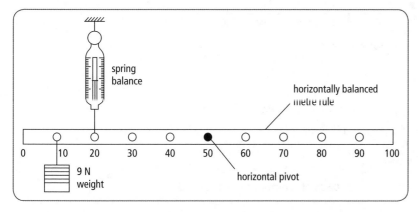

Anticlockwise moment due to the 9 N weight $= 9 \times 0.4 = 3.6$ Nm
Reading on the spring balance $= 12$ N
Moment due to the force of the spring balance $= 12 \times 0.3 = 3.6$ Nm

Conclusion: in equilibrium, clockwise moment $=$ anticlockwise moment

Examination style questions

1. a. State the two factors on which the turning effect of a force depends.

b. Forces F_1 and F_2 are applied vertically downwards at the ends of a beam resting on a pivot P. The beam has weight W. The beam is shown in the diagram below.

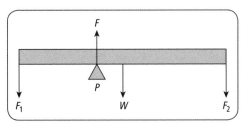

i) Complete the statements about the two requirements for the beam to be in equilibrium.

1. There must be no resultant ....

2. There must be no resultant ....

ii) The beam is in equilibrium. F is the force exerted on the beam by the pivot P. Complete the following equation about the forces on the beam.

$F = $ ....

iii) Which one of the four forces on the beam does **not** exert a moment about P?

CIE 0625 November '06 Paper 2 Q5

2. The diagram below shows apparatus for investigating moments of forces.

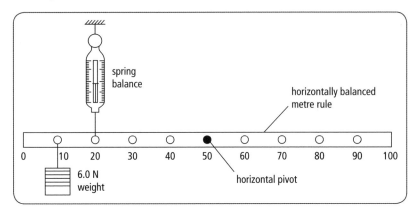

The uniform metre rule shown is in equilibrium.

a. Write down two conditions for the metre rule to be in equilibrium.

b. Show that the value of the reading on the spring balance is 8.0 N.

c. The weight of the uniform metre rule is 1.5 N. Calculate the force exerted by the pivot on the metre rule and state its direction.

CIE 0625 November '05 Paper 3 Q2

Practical question

The IGCSE class is determining the weight of a metre rule.

Below is a diagram of the apparatus

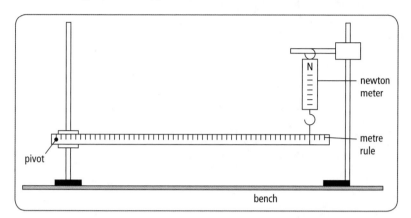

A metre rule is supported at one end by a pivot through the 1.0 cm mark. The other end is supported at the 91.0 cm mark by a newton meter hanging from a clamp.

1. Describe how you would check that the metre rule is horizontal. You may draw a diagram if you wish.

2. The students record the force F shown on the newton meter and the distance d from the pivot to the 91 cm mark. They then repeat the experiment several times using a range of values of the distance d. The readings are shown in the table.

F/N	d/m	$\frac{1}{d} / \frac{1}{\text{m}}$
0.74	0.900	
0.78	0.850	
0.81	0.800	
0.86	0.750	
0.92	0.700	

Copy the table. Calculate and record on your table the values of $1/d$

3. a. On graph paper, plot a graph of F/N (y-axis) against $\frac{1}{d} / \frac{1}{\text{m}}$ (x-axis). Start the y-axis at 0.7 and the x-axis at 1.0
   b. Draw the line of best fit on your graph.
   c. Determine the gradient G of the line.

4. Calculate the weight of the metre rule using the equation
$$W = \frac{G}{k} \quad \text{where } k = 0.490 \text{ m.}$$

Centre of mass

The centre of mass of an object is the point on the object where the mass can be considered to be concentrated and hence where the weight of the object can be considered to act.

The centre of mass of a very thin object (a lamina) can be found by experiment:

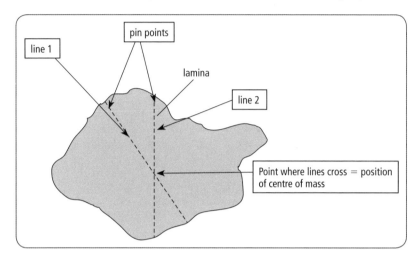

Push a pin through a point on the edge of the lamina and allow it to swing freely. Use a plumb line (a small mass on a piece of string) to mark a vertical line from the pin point across the lamina. Repeat for a second point on the edge of the lamina. Where the two lines cross is the position of the centre of mass.

The position of the centre of mass affects the **stability** of an object

For example:

If object 1 is tilted through a small angle, the **weight will act outside the base**. There will be a net moment on object 1 that will cause it to fall over. If object 2 is tilted through a small angle the weight will **still act inside the base**. There will be a net moment on object 2 that will cause it to go back to its original position and it will not fall over.

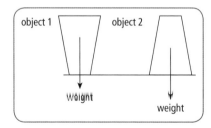

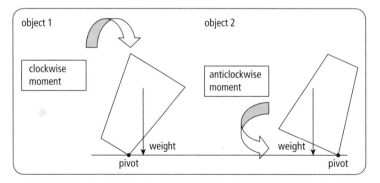

Examination style questions

1. A piece of stiff cardboard is stuck to a plank of wood by means of two sticky-tape "hinges". This is shown below.

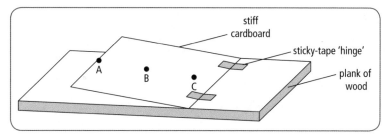

a. Initially, the cardboard is flat on the plank of wood. A box of matches is placed on it. The cardboard is then slowly raised at the left hand edge, as shown below. State the condition for the box of matches to fall over.

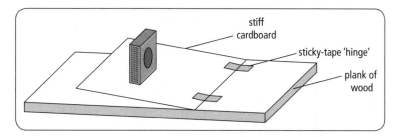

b. The box of matches is opened, as shown below. The procedure in (a) is repeated.

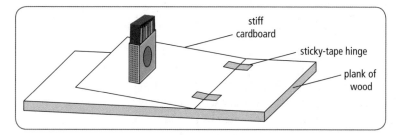

  i)   Complete the sentence below, using either the words "greater than" or "the same as" or "less than".

    In (b), the angle through which the cardboard can be lifted before the box of matches falls is .... the angle before the box of matches falls in (a).

  ii) Give a reason for your answer to (i).

Adapted from CIE 0625 June '07 Paper 2 Q3

2. a. A light vertical triangular piece of rigid plastic PQR is pivoted at corner P. A horizontal 5N force acts at Q, as shown in below.

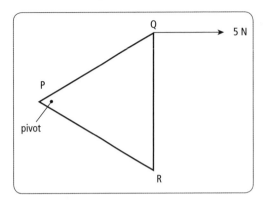

Describe what, if anything, will happen to the piece of plastic.

b. On another occasion, two horizontal 5N forces act on the piece of plastic, as shown in below.

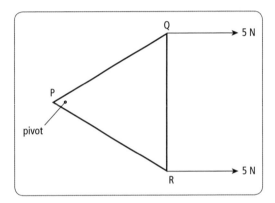

i) Describe what, if anything, will happen to the piece of plastic.

ii) Copy the diagram above and mark the force that the pivot exerts on the piece of plastic. Show the direction of the force by means of an arrow and write the magnitude of the force next to the arrow.

CIE 0625 June '05 Paper 2 Q3

Scalars and vectors

A **scalar** quantity has **size** only.

A **vector** quantity has **size** and **direction**.

Examples of scalar quantities	Examples of vector quantities
Mass	Velocity
Energy	Acceleration
Time	Force

Resultants

To calculate the resultant (overall) force on a point acted on by two forces, F_1 and F_2 you can draw a scale diagram.

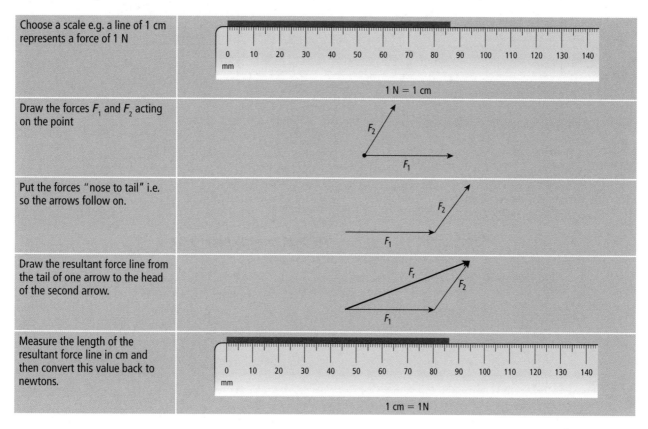

Choose a scale e.g. a line of 1 cm represents a force of 1 N	1 N = 1 cm
Draw the forces F_1 and F_2 acting on the point	
Put the forces "nose to tail" i.e. so the arrows follow on.	
Draw the resultant force line from the tail of one arrow to the head of the second arrow.	
Measure the length of the resultant force line in cm and then convert this value back to newtons.	1 cm = 1N

Examination style question

1. A student sets up the apparatus shown below in order to find the resultant of the two tensions T_1 and T_2 acting at P. When the tensions T_1, T_2 and T_3 are balanced, the angles between T_1 and the vertical and T_2 and the vertical are as marked on the diagram.

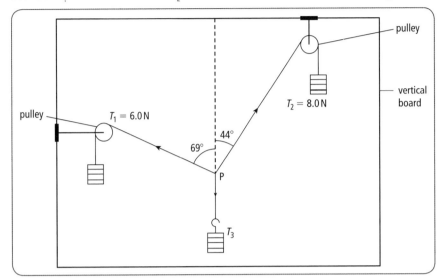

Draw a scale diagram of the forces T_1 and T_2. Use the diagram to find the resultant of the two forces.

State:
a. The value of the resultant
b. The direction of the resultant
c. The scale used in the drawing

CIE 0625 June '06 Paper 3 Q2

1.6 Energy, work and power

Energy

An object may have energy because it is moving or because of its position. Energy can be **transferred** from one place to another, **transformed** from one type to another or **stored**. The unit of energy is the **joule (J)**.

Types of energy

Gravitational potential	The energy gained as an object is moved away from the Earth e.g. a book being lifted onto a shelf
Kinetic	The energy an object has due to its movement e.g. a person running
Chemical	Stored energy that can be released in a chemical reaction e.g. a battery, fuel such as coal
Strain	The energy stored when an object changes shape e.g. a stretched rubber band
Electric	The energy carried by an electric current
Sound	The energy carried by a sound wave
Internal energy	The total kinetic and potential energies of all of the particles in an object
Thermal (heat) energy	The energy released when the temperature of a hot object decreases due to a decrease in its internal energy
Nuclear	Stored energy that can be released in a nuclear reaction e.g. energy stored in the sun.
Light	Energy given off, for example, by very hot objects

Heat energy can be **transferred** from a hot object to a cooler one.

Kinetic energy can be **transferred** from one car to another in a collision.

Energy transformations

In an energy transformation, energy is converted from one type to another. For example:

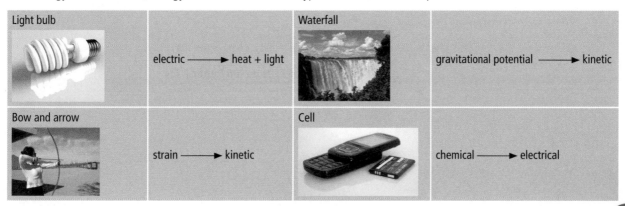

Light bulb: electric ⟶ heat + light

Waterfall: gravitational potential ⟶ kinetic

Bow and arrow: strain ⟶ kinetic

Cell: chemical ⟶ electrical

Principle of conservation of energy

Energy cannot be created or destroyed. It is transformed from one form to another.

Example

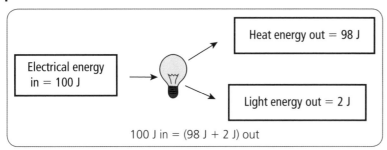

Heat energy out = 98 J

Electrical energy in = 100 J

Light energy out = 2 J

100 J in = (98 J + 2 J) out

Examination style questions

1. A piece of fruit is falling from a tree.

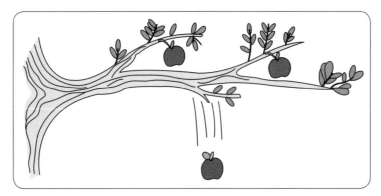

The list below contains the names of some different forms of energy.

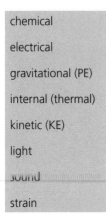

chemical

electrical

gravitational (PE)

internal (thermal)

kinetic (KE)

light

sound

strain

a. Which **four** forms of energy are possessed by the falling fruit?

b. Which form of energy increases as the fruit falls?

c. Which form of energy decreases as the fruit falls?

d. Which form of energy is stored in the body of a person as a result of eating the fruit?

CIE 0625 June '06 Paper 2 Q4

2. A child is sitting on an oscillating swing, as shown below. At the top of the oscillation, the child and swing are momentarily at rest.

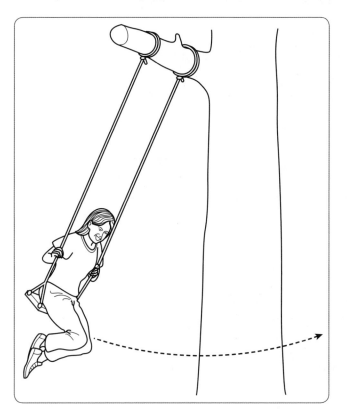

i) Use the names of appropriate types of energy to complete the following word equation.

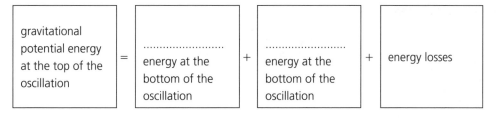

| gravitational potential energy at the top of the oscillation | = | .... energy at the bottom of the oscillation | + | .... energy at the bottom of the oscillation | + | energy losses |

ii) The child continues to sit still on the swing. The amplitude of the oscillations slowly decreases.
Explain why this happens.

CIE 0625 November '06 Paper 2 Q4b and c

Energy resources

Energy resources are used to produce electrical energy from other forms of energy.

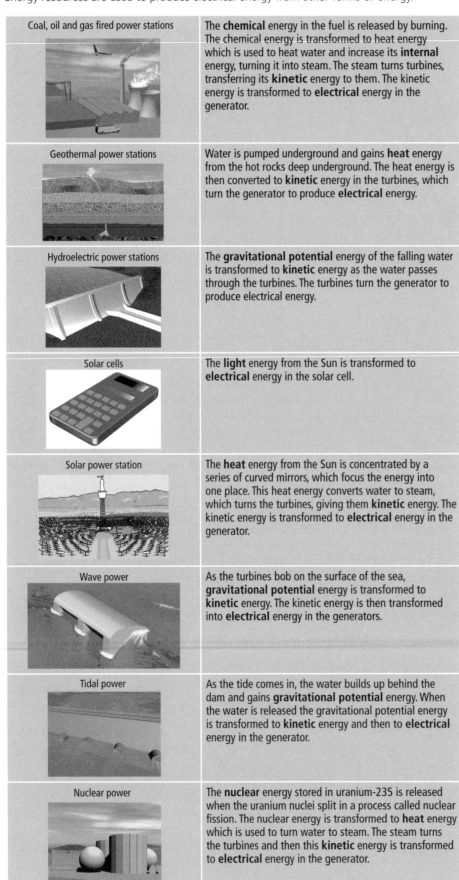

Coal, oil and gas fired power stations	The **chemical** energy in the fuel is released by burning. The chemical energy is transformed to heat energy which is used to heat water and increase its **internal** energy, turning it into steam. The steam turns turbines, transferring its **kinetic** energy to them. The kinetic energy is transformed to **electrical** energy in the generator.
Geothermal power stations	Water is pumped underground and gains **heat** energy from the hot rocks deep underground. The heat energy is then converted to **kinetic** energy in the turbines, which turn the generator to produce **electrical** energy.
Hydroelectric power stations	The **gravitational potential** energy of the falling water is transformed to **kinetic** energy as the water passes through the turbines. The turbines turn the generator to produce electrical energy.
Solar cells	The **light** energy from the Sun is transformed to **electrical** energy in the solar cell.
Solar power station	The **heat** energy from the Sun is concentrated by a series of curved mirrors, which focus the energy into one place. This heat energy converts water to steam, which turns the turbines, giving them **kinetic** energy. The kinetic energy is transformed to **electrical** energy in the generator.
Wave power	As the turbines bob on the surface of the sea, **gravitational potential** energy is transformed to **kinetic** energy. The kinetic energy is then transformed into **electrical** energy in the generators.
Tidal power	As the tide comes in, the water builds up behind the dam and gains **gravitational potential** energy. When the water is released the gravitational potential energy is transformed to **kinetic** energy and then to **electrical** energy in the generator.
Nuclear power	The **nuclear** energy stored in uranium-235 is released when the uranium nuclei split in a process called nuclear fission. The nuclear energy is transformed to **heat** energy which is used to turn water to steam. The steam turns the turbines and then this **kinetic** energy is transformed to **electrical** energy in the generator.

Nuclear fusion

The process of nuclear fusion is carried out in the Sun. Hydrogen nuclei collide at great speed in the Sun and fuse together to form helium nuclei. This releases energy in the form of heat and light.

Examination style question

1. The diagram below represents a hydroelectric system for generating electricity.

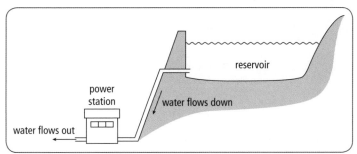

Answer the following questions, using words from this list.

chemical

electrical

gravitational

internal (heat)

kinetic

light

nuclear

sound

strain

   a.   What sort of energy, possessed by the water in the reservoir, is the main source of energy for this system?
   b.   When the water flows down the pipe, it is moving. What sort of energy does it possess because of this movement?
   c.   The water makes the turbines in the power station rotate. What sort of energy do the turbines possess because of their rotation?
   d.   What sort of energy does the power station generate?
   e.   None of the energy transfer processes is perfect. In what form is most of the wasted energy released?

CIE 0625 June '05 Paper 2 Q4

Work and power

The **work done** in **joules** by a force acting on an object = force × distance moved by the object in the direction of the force.

$$\text{work done (J)} = \text{force (N)} \times \text{distance (m)}$$
$$W = F \times d$$

The **power** in **watts** is the work done per second or the energy transformed per second.

$$\text{power (W)} = \frac{\text{energy (J)}}{\text{time (s)}}$$
$$P = \frac{E}{t}$$

Worked examples

1. A car engine produces a force of 2000 N while accelerating the car through a distance of 200 m in a time of 10 s.
   a. What is the work done on the car by the engine force?
   b. What is the power developed by the engine?

Answers

a. $W = F \times d$
   $= 2000 \times 200$
   $= 400\,000$ J
   $= 400$ kJ

b. $P = \frac{E}{t}$
   $= \frac{400\,000}{10}$
   $= 40\,000$ W
   $= 40$ kW

Kinetic energy

Kinetic energy can be calculated from the formula:
$$KE = \frac{1}{2}mv^2$$
where m = mass in kg; v = velocity in m/s

When an object is lifted higher above the Earth's surface work must be done.
Since work = force × distance
and force = weight of the object
work done = weight × height lifted
where weight = mass × gravitational field = mg.

Gravitational potential energy

It follows that the change in **gravitational potential energy** (the work done in lifting the object) is given by the formula:
$$PE = mgh$$
where m = mass in kg; g = acceleration due to gravity 10m/s²;
      h = change in height in m

Worked examples

1. A car of mass 1000 kg is travelling at a velocity of 20 m/s. Calculate its kinetic energy.

2. Calculate the change in potential energy of a 70 kg parachutist as she falls through a height of 100 m.

3. A ball of mass 0.5 kg is dropped from rest at a height of 5 m above the ground. Find its velocity when it hits the ground.

Answers

1. $KE = \frac{1}{2}mv^2$

$= \frac{1}{2} \times 1000 \times 20^2$

$= 200\ 000$ J

$= 200$ kJ

2. $PE = mgh$

$= 70 \times 10 \times 100$

$= 70\ 000$ J

$= 70$ kJ

3. $PE = mgh$

$= 0.5 \times 10 \times 5$

$= 25$ J

loss of PE = gain of KE

$v = \sqrt{(2KE/m)}$

$= \sqrt{(2 \times 25/0.5)}$

$= 10$ m/s

Examination style questions

1. An electric pump is used to raise water from a well, as shown below.

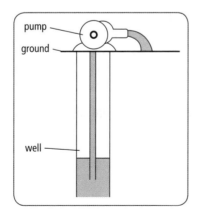

a. The pump does work in raising the water. State an equation that could be used to calculate the work done in raising the water.

b. The water is raised through a vertical distance of 8.0 m. The weight of water raised in 5.0 s is 100 N.
   i)   Calculate the work done in raising the water in this time.
   ii) Calculate the power the pump uses to raise the water.
   iii) The energy transferred by the pump to the water is greater than your answer to (i).
        Suggest what the additional energy is used for.

CIE 0625 June '06 Paper 3 Q3

2. A student wishes to work out how much power she uses to lift her body when climbing a flight of stairs.
Her body mass is 50 kg and the vertical height of the stairs is 4.0 m. She takes 20 s to walk up the stairs.
a. Calculate
   i)   the work done in raising her body mass as she climbs the stairs,
   ii) the output power she develops when raising her body mass.
b. At the top of the stairs she has gravitational potential energy. Describe the energy transformations taking place as she walks back down the stairs and stops at the bottom.

CIE 0625 June '07 Paper 3 Q3

3. The diagram below shows water falling over a dam.

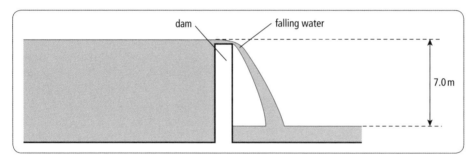

a. The vertical height that the water falls is 7.0 m.
   Calculate the potential energy lost by 1.0 kg of water during the fall.
b. Assuming all this potential energy loss is changed to kinetic energy of the water, calculate the speed of the water, in the vertical direction, at the end of the fall.
c. The vertical speed of the water is less than that calculated in b. Suggest one reason for this.

CIE 0625 November '06 Paper 3 Q3

1.7 Pressure

The **pressure** on a surface due to a force is the force on 1 m² of the surface.

$$\text{pressure (N/m}^2) = \frac{\text{force (N)}}{\text{area (m}^2)}$$

$$p = \frac{F}{A}$$

The unit of pressure, N/m² is also known as the **pascal (Pa)**

Worked examples

1. A force of 10 kN acts on the surface of a liquid, of area 0.08 m². What is the pressure on the surface of the liquid?

2. A person of weight 600 N exerts a pressure of 200 kPa on the ground. What is the area of their feet?

3. The area of a dog's paw is 10 cm². The pressure under the paw is 50 kPa when it exerts half of its body weight on the paw. What is its weight?

Answers

1. $p = \dfrac{F}{A}$

$= \dfrac{10\ 000}{0.08}$

$= 125\ 000$ Pa

$= 125$ kPa

2. $A = \dfrac{F}{p}$

$= \dfrac{600}{200\ 000}$

$= 0.003$ m²

3. $F = p \times A$

$= 50\ 000 \times (10 \div 10\ 000)$

$= 50$ N

Total weight $= 2 \times 50 = 100$ N

Note: In question 1, 1 kN = 1000 N

In question 3, area in cm² is converted to area in m² by dividing by (100 × 100) ie by 10 000

Atmospheric (air) pressure can be measured with a **barometer**.

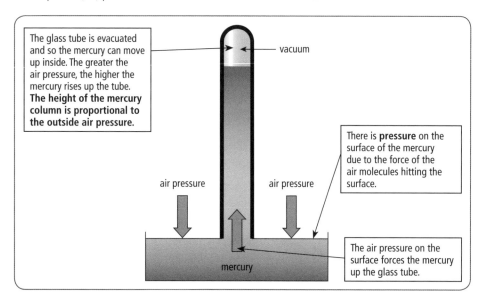

The glass tube is evacuated and so the mercury can move up inside. The greater the air pressure, the higher the mercury rises up the tube. **The height of the mercury column is proportional to the outside air pressure.**

vacuum

air pressure

air pressure

There is **pressure** on the surface of the mercury due to the force of the air molecules hitting the surface.

The air pressure on the surface forces the mercury up the glass tube.

mercury

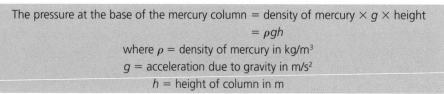

The pressure at the base of the mercury column = density of mercury $\times$ g $\times$ height
$= \rho g h$
where ρ = density of mercury in kg/m³
g = acceleration due to gravity in m/s²
h = height of column in m

The pressure below the surface of all liquids increases proportionally to depth. For example, the formula above can be used to calculate the pressure under the surface of the sea.

A **manometer** can be used to measure the pressure of a gas:

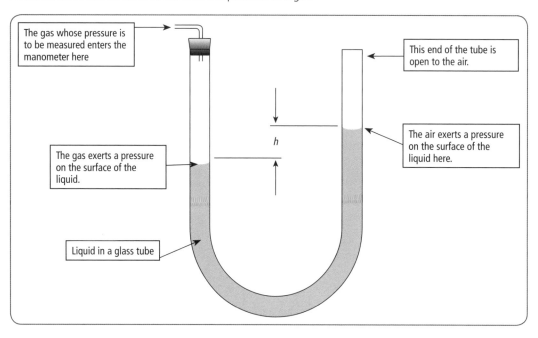

The gas whose pressure is to be measured enters the manometer here

This end of the tube is open to the air.

The gas exerts a pressure on the surface of the liquid.

h

The air exerts a pressure on the surface of the liquid here.

Liquid in a glass tube

The pressure due to the gas = atmospheric pressure + $\rho g h$
Where ρ = density of liquid in kg/m³; h = difference in height of two liquid surfaces in m

Examination style questions

1. a. The diagram below shows two examples of footwear being worn by people of equal weight at a Winter Olympics competition.
   Which footwear creates the greatest pressure below it, and why?

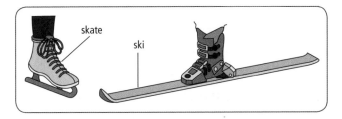

   b. Drivers of high-sided vehicles, like the one below, are sometimes warned not to drive when it is very windy.

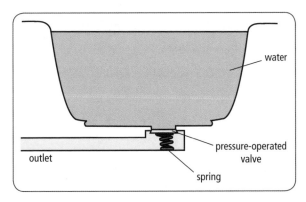

   Suggest why they receive this warning.

CIE 0625 November '05 Paper 2 Q3

2. The diagram below shows a pond that is kept at a constant depth by a pressure-operated valve in the base.

   a. The pond is kept at a depth of 2.0 m. The density of water is 1000 kg/m³.
      Calculate the water pressure on the valve.
   b. The force required to open the valve is 50 N. The valve will open when the water depth reaches 2.0 m. Calculate the area of the valve.
   c. The water supply is turned off and the valve is held open so that water drains out through the valve. State the energy changes of the water that occur as the depth of the water drops from 2.0 m to zero.

CIE 0625 November '05 Paper 3 Q3

Summary questions on unit 1

1. Fill in the blanks

The speed of an object can be calculated by working out _____

divided by _____.

The unit of speed is _____. Speed is equal to the _____ of a

distance–time graph. Velocity is equal to speed in a certain _____ and

it has the same _____. The acceleration of an object is equal to

the _____ _____ _____ of velocity and has units of _____.

Acceleration can be calculated from the _____ of a velocity–time graph. The

area under a velocity–time graph is equal to the _____.

2. Calculate the speed of an object that travels 10 m in 5 s.

3. The initial velocity of a car is 10 m/s. It reaches a velocity of 25 m/s in 5 s. What is its acceleration?

4. What is the gradient of a distance–time equal to?

5. What is the area under a velocity–time graph equal to?

6. What is the difference between mass and weight?

7. A pebble of mass 100 g is placed in a measuring cylinder containing 50 ml of water. The water level rises to 75 ml. What is the density of the pebble?

8. Fill in the blanks

A force can stretch an object, or change its _____. If small forces are applied to

a spring, the extension produced by a load is _____ _____ to the force

applied. This is true up to the limit of _____. A force can change the velocity of

an object by causing it to _____. The acceleration is _____ if the force

on the object is doubled and the mass stays constant. The net or overall force on an

object is called the _____ force. When forces act in opposite directions

they are _____ to find the resultant force. When forces act in the same

direction, they are _____ to find the resultant force. A force can change

the _____ in which an object is travelling without changing its _____.

This happens when the force is _____ to the direction in which the object is

travelling and it causes the object to move in a _____.

9. A force of 10 N produces an extension of 20 cm. What extension would be produced by a force of 2.5 N?

10. Fill in the blanks

The mass of an object is measured in _____. The weight of an object is

a type of _____ and is measured in _____. The weight can be

calculated by multiplying the _____ by the _____ _____

_____. The density of an object is equal to its mass divided by its

_____. The unit of density is _____ or _____.

11. Match the devices to the energy transformations

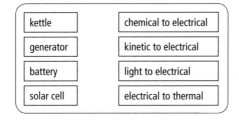

kettle	chemical to electrical
generator	kinetic to electrical
battery	light to electrical
solar cell	electrical to thermal

12. Calculate the kinetic energy of a ball of mass 0.2 kg thrown at a speed of 7 m/s.

13. Calculate the increase in gravitational potential energy of a boy of mass 50 kg
climbing a flight of stairs of height 5 m.

14. Calculate the distance travelled by an object if the work done on it by a force of 20 N
is 100 J.

15. What is the power of a kettle that transforms 100 J of electrical energy to internal
energy in 0.5 s?

16. What is the pressure 3 m under the surface of the sea? Sea water has a density of
1200 kg/m³.

17. Calculate the pressure under a block of mass 0.3 kg when resting on an area of 5 cm².

Examination style questions on unit 1

1. A solid plastic sphere falls towards the Earth.
Below is the speed-time graph of the fall up to the point where the sphere hits the Earth's surface.

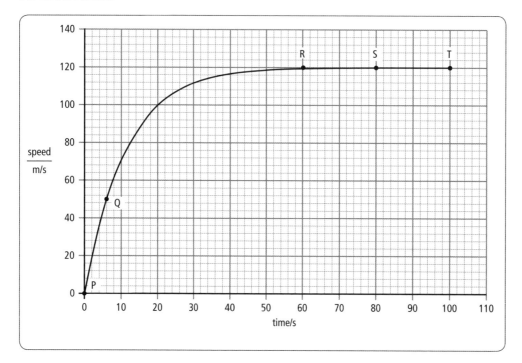

a. Describe in detail the motion of the sphere shown by the graph.
b. Copy the diagram below and, draw arrows to show the directions of the forces acting on the sphere when it is at the position shown by point S on the graph. Label your arrows with the names of the forces.

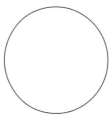

c. Explain why the sphere is moving with constant speed at S.
d. Use the graph to calculate the approximate distance that the sphere falls
    i)   between R and T,
    ii) between P and Q.

CIE 0625 June '05 Paper 3 Q1

2. The diagram below shows a simple pendulum that swings backwards and forwards between P and Q.

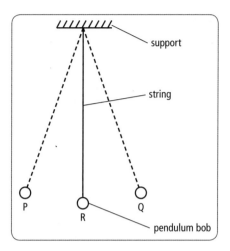

a. The time taken for the pendulum to swing from P to Q is approximately 0.5 s. Describe how you would determine this time as accurately as possible.

b. i) State the two vertical forces acting on the pendulum bob when it is at position R.

   ii) The pendulum bob moves along the arc of a circle. State the direction of the resultant of the two forces in (i).

c. The mass of the bob is 0.2 kg. During the swing it moves so that P is 0.05 m higher than R.

   Calculate the increase in potential energy of the pendulum bob between R and P.

CIE 0625 June '05 Paper 3 Q2

3. A bus travels from one bus stop to the next. The journey has three distinct parts. Stated in order they are

   uniform acceleration from rest for 8.0 s,

   uniform speed for 12 s,

   non-uniform deceleration for 5.0 s.

The graph below shows only the deceleration of the bus.

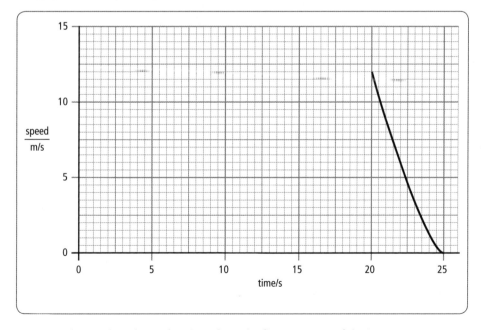

a. Copy the graph and complete it to show the first two parts of the journey.

b. Calculate the acceleration of the bus 4.0 s after leaving the first bus stop.

c. Use the graph to estimate the distance the bus travels between 20 s and 25 s.
d. On leaving the second bus stop, the uniform acceleration of the bus is 1.2 m / s².
The mass of the bus and passengers is 4000 kg.
Calculate the accelerating force that acts on the bus.
e. The acceleration of the bus from the second bus stop is less than that from the first bus stop.
Suggest two reasons for this.

CIE 0625 June '06 Paper 3 Q1

4. The diagram below shows a model car moving clockwise around a horizontal circular track.

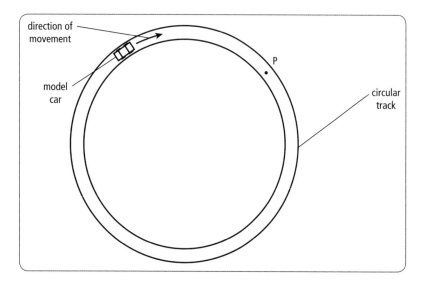

a. A force acts on the car to keep it moving in a circle.
   i) Copy the diagram and draw an arrow to show the direction of this force.
   ii) The speed of the car increases. State what happens to the magnitude of this force.
b. i) The car travels too quickly and leaves the track at P. On your diagram, draw an arrow to show the direction of travel after it has left the track.
   ii) In terms of the forces acting on the car, suggest why it left the track at P.
c. The car, starting from rest, completes one lap of the track in 10 s. Its motion is shown graphically in below.

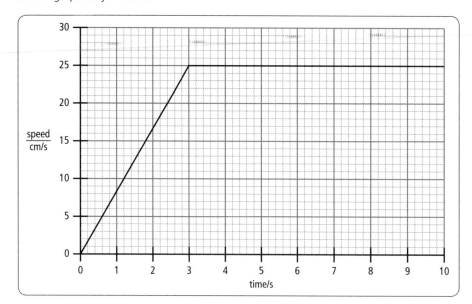

    i)   Describe the motion between 3.0 s and 10.0 s after the car has started.

    ii)   Use the diagram at the beginning of the question to calculate the circumference of the track.

    iii)   Calculate the increase in speed per second during the time 0 to 3.0 s.

CIE 0625 June '07 Paper 3 Q1

5. The diagram below shows a steam safety valve. When the pressure gets too high, the steam lifts the weight W and allows steam to escape.

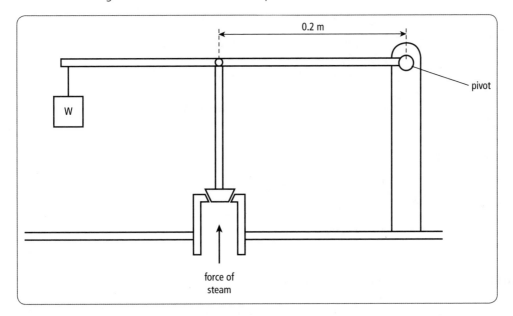

  a.   Explain, in terms of moments of forces, how the valve works.

  b.   The moment of weight W about the pivot is 12 N m. The perpendicular distance of the line of action of the force of the steam on the valve from the pivot is 0.2 m. The area of the piston is 0.0003 m².

    Calculate

    i)   the minimum steam force needed for the steam to escape,

    ii)   the minimum steam pressure for the steam to escape.

CIE 0625 June '07 Paper 3 Q2

2 Thermal physics

2.1 Kinetic theory

KEY IDEAS

✓ The arrangement of particles varies in solids, liquids and gases
✓ The arrangement of particles affects the physical properties of a substance
✓ The kinetic energy of the particles is dependent on the temperature of the substance
✓ The particles in gases and liquids are in constant random motion
✓ When the particles of a gas collide with the walls of its container they exert a pressure
✓ The pressure due to a gas at constant temperature is inversely proportional to its volume
✓ The highest energy particles escape the surface of a liquid due to evaporation, at temperatures below the boiling point

A molecule is one or more atoms (of the same or different elements, joined together by chemical bonds) capable of existing as a unit.

Molecular structure of solids liquids and gases

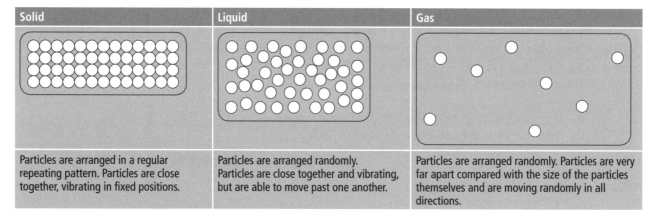

Solid	Liquid	Gas
Particles are arranged in a regular repeating pattern. Particles are close together, vibrating in fixed positions.	Particles are arranged randomly. Particles are close together and vibrating, but are able to move past one another.	Particles are arranged randomly. Particles are very far apart compared with the size of the particles themselves and are moving randomly in all directions.

Physical properties of solids liquids and gases

Solid	Liquid	Gas
Very difficult to compress	Difficult to compress	Easy to compress
Cannot flow	Can flow	Can flow
Keeps its shape	Takes the shape of the bottom of the container	Fills the entire container

Solids

In a solid the particles are close together with strong bonds (attractive forces) between them. It is difficult to separate the particles and as a result solids are hard to break. As the particles are already tightly packed, they cannot be pushed closer together and so solids are **incompressible**.

Liquids

In a liquid, the particles are close together with weak bonds (attractive forces) between them. It is easy for the particles to move past one another and as a result liquids can be poured and can flow. As the particles are already close together, they cannot be pushed much closer together and so liquids are **incompressible**.

Gases

In a gas, the particles are very far apart with negligible (practically zero) forces between them. It is easy for the particles to move past one another and the gas can flow. The particles can easily be pushed closer together and so gases are **compressible**.

Substances that can flow are called **fluids**. All liquids and gases are fluids.

Change in motion of molecules in a gas with temperature

As the temperature of a gas increases, the average kinetic energy of the particles increases and so the average speed of the particles increases.

Pressure due to a gas

The gas particles are moving fast in all directions. If the gas is placed in a sealed container, when the particles hit the sides of the container, they rebound and so exert a force on the walls of the container. Hence there is a pressure on the side of the container equal to the force exerted divided by the area of the side. ($p = F/A$, see section 1.7). If the temperature of the gas increases, the average speed of the particles and the number of collisions per second increases. Hence the force exerted on the container by the particles is greater and so the pressure increases.

Brownian motion: the movement of small particles in a gas or a liquid

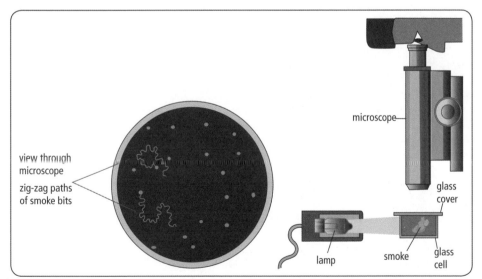

view through microscope

zig-zag paths of smoke bits

microscope

glass cover

lamp   smoke   glass cell

If smoke is introduced into a glass cell and observed through a microscope, the smoke particles can clearly be seen. The smoke particles appear to be in **random motion**. This is because they are constantly being hit by the other rapidly moving particles, in the air, in the glass cell.

This is evidence that the particles in air are in constant random motion. It shows that massive particles may be moved by light, fast-moving particles.

Boyle's Law: the variation of volume of a gas with applied pressure

Experimental arrangement

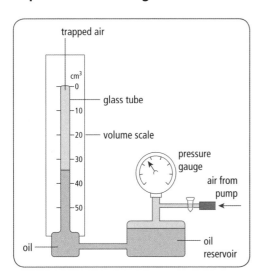

Experimental procedure

The pressure on a fixed mass of gas is varied by using a foot pump and measured using a pressure gauge at constant temperature. The volume of the gas is recorded as the pressure is increased.

Results

As the graph shows, when the pressure on the gas is increased, the volume of the gas decreases. If the pressure is doubled, the volume of the gas halves. The volume is **inversely proportional** to the pressure applied.

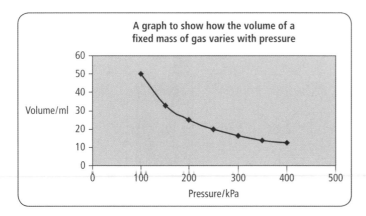

▲ A graph to show how the volume of a fixed mass of gas varies with pressure

Boyle's Law

For a fixed mass of gas at constant temperature, the volume is inversely proportional to the applied pressure.

$$pV = \text{constant}$$

Evaporation

Evaporation is the transformation of a liquid into a gas at a temperature **below the boiling point of the liquid.** The liquid particles have kinetic energy and are moving past one another but some particles have more kinetic energy than the others. These faster particles may be moving fast enough to overcome the force of attraction between them and the other particles. The fast moving particles escape from the **surface** of the liquid and form a gas.

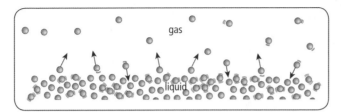

The rate of evaporation can be increased by:

*   increasing the temperature (e.g. put the liquid in a warmer room)
*   increasing the surface area (e.g. put the liquid in a wider container)
*   increasing the air flow over the surface of the liquid (e.g. blow on the surface).

Examination style questions

1. The following diagram shows a plastic bottle containing only air. The bottle is sealed by the cap.

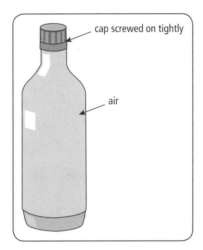

    a.   Describe the motion of the molecules in the air inside the bottle.
    b.   Describe and explain what happens to the pressure in the bottle as the temperature of the air increases.

Adapted from CIE 0625 June '05 Paper 2 Q5a

2. An IGCSE student notices a puddle of water on the pavement as he leaves home on a warm day. When he returns one hour later, the puddle is only half as wide as before.
    a.   State the name of the process causing the decrease in the size of the puddle.
    b.   Explain how the process leads to the observed effect.

Adapted from CIE 0625 June '05 Paper 2 Q5b

3. The apparatus shown below is set up in a laboratory during a morning science lesson.

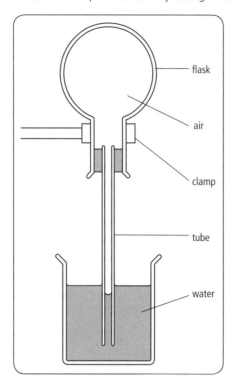

Later in the day, the room temperature is higher than in the morning.

a.   What change is observed in the apparatus?

b.   Explain why this change happens

c.   Suggest one disadvantage of using this apparatus to measure temperature

CIE 0625 November '05 Paper 2 Q5

4. a.   The diagram below shows the paths of a few molecules in air and a single dust particle. The actual molecules are too small to show on the diagram.

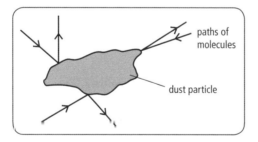

Explain why the dust particle undergoes small random movements.

b.   The diagram below shows the paths of a few molecules leaving the surface of a liquid. The liquid is below its boiling point.

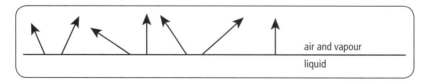

i)   State which liquid molecules are most likely to leave the surface.

ii)   Explain your answer to (i).

CIE 0625 June '05 Paper 3 Q5

5 The diagram below shows a way of indicating the positions and direction of movement of some molecules in a gas at one instant.

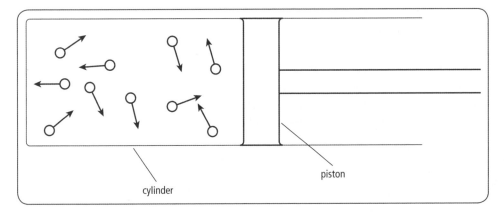

cylinder

piston

a. i) Describe the movement of the molecules.
   ii) Explain how the molecules exert a pressure on the container walls.
b. When the gas in the cylinder is heated, it pushes the piston further out of the cylinder. State what happens to
   i) the average spacing of the molecules,
   ii) the average speed of the molecules.
c. The gas shown in the diagram is changed into a liquid and then into a solid by cooling. Compare the gaseous and solid states in terms of
   i) the movement of the molecules,
   ii) the average separation of the molecules.

CIE 0625 November '05 Paper 3 Q5

Practical question

An IGCSE student sets up the following experiment by wrapping the bulbs of three alcohol-in-glass thermometers with the same thickness of cotton wool. Thermometer 1 has dry cotton wool, thermometer 2 has cotton wool soaked in water and thermometer 3 has cotton wool soaked in acetone. She then investigates how the temperature of each thermometer varies with time.

▲ Thermometer 1 dry   ▲ Thermometer 2 water   ▲ Thermometer 3 acetone

The results of the experiment are shown in the table below:

Time/min	Temperature 1/°C	Temperature 2/°C	Temperature 3/°C
1	24	24	24
2	23	21	20
3	23	19	17
4	23	18	16
5	23	18	15

1. Compare and explain the trend in temperature for thermometer 1 and thermometer 2.

2. Compare and explain the results for thermometer 2 and thermometer 3.

2.2 Thermal properties

Thermal expansion of a solid

When solids are heated, they expand only a little, usually too little to see with the naked eye. However, in large structures such as bridges, buildings and railway tracks, this expansion could cause problems when there are large changes in temperature. To avoid this, railway tracks have small gaps between them to allow for expansion. If no gap was left, when the tracks expanded on a hot day they would be forced against each other and the track would bend and buckle.

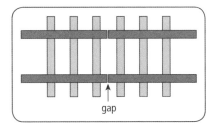

▲ Railway tracks on a cold day   ▲ Railway tracks on a very hot day

The expansion of metals can be used in a **bimetallic strip**, shown below.

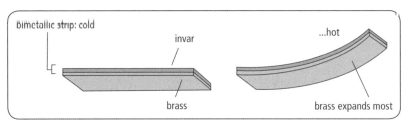

When the strip is heated, one metal expands more than the other and so the strip bends. The higher the temperature, the more the strip bends. In a **thermostat**, when the temperature rises, the bimetallic strip bends and the contacts separate, switching off the current to the heater. When the temperature falls, the strip goes back to its original position and the heater is switched on again. The temperature at which this occurs can be altered using the adjustable control knob.

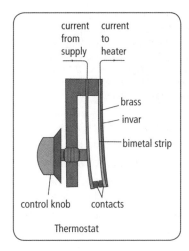

▲ Thermostat

Thermal expansion of a liquid

Liquids expand more than solids when heated, enough for the effect to be visible. This expansion is used in liquid-in-glass thermometers.

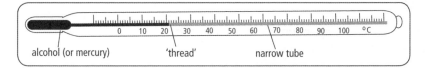

When the alcohol in the thermometer is heated by placing the bulb in a hot liquid, for example, the alcohol heats up and expands. This forces the alcohol up the narrow tube, so the thermometer gives a higher temperature reading.

Thermal expansion of a gas

Gases expand more than liquids when heated. When the temperature of a gas is increased, the particles gain kinetic energy and move around faster. This means they take up more space and so, if the gas is free to expand, it expands.

Measurement of temperature

The **Celsius** scale of temperature is defined by its two fixed points: the freezing and boiling point of water. These are taken to be 0 °C and 100 °C (in standard atmospheric conditions).

The **Kelvin** scale of temperature is defined by the vibrations of particles. At **absolute zero**, (0 kelvin or **0 K**) particles have zero kinetic energy and no longer vibrate. Compared to the Celsius scale, this occurs at **−273 °C**. The two scales have the same sized degree, 1 °C is 274K so 2 °C is 275K.

Kelvin temperature/K = Celsius temperature/°C + 273

Examples of thermometers

Liquid-in-glass	See expansion of liquids, above
Thermistor	A thermistor is an electrical component whose resistance decreases with temperature. It allows more current to flow when the temperature increases. The current reading is converted to a temperature reading on a digital meter.
Thermocouple	A thermocouple is made from two different types of metal, which are joined together to form two junctions. When there is a difference in temperature between the junctions, a small potential difference is created that depends on the temperature difference between the two junctions. The "hot junction" is placed in the substance whose temperature is to be measured. The potential difference between the hot and the cold junctions is then measured. The greater the temperature of the hot junction, the greater the potential difference. The potential difference is then converted to a temperature reading.

All thermometers calibrated on the Celsius scale agree at the fixed points of 0 °C and 100 °C but not necessarily at temperatures in between these values. This is because the property of the material that is being used to measure temperature may not vary linearly. This affects the accuracy of the thermometer. The precision, or sensitivity, of the thermometer depends on how finely it is calibrated. For example, in an alcohol-in-glass thermometer, the divisions usually allow temperature to be measured to the nearest 1°C. Digital meters attached to a thermocouple or thermistor may give temperature to the nearest 0.1°C.

The range of a liquid-in-glass thermometer is limited by the freezing and boiling points of the liquid. A frozen thermometer will not work and if the liquid boils it will explode! Thermistor and thermocouple thermometers may have greater ranges but are not always as convenient as a liquid-in-glass thermometer.

Examination style questions

1. A manufacturer of liquid-in-glass thermometers changes the design in order to meet new requirements. Describe the changes that could be made to increase:
  a. the range of the thermometer
  b. the sensitivity of the thermometer

Adapted from CIE 0625 June '06 Paper 3 Q5b

2. The diagram below shows a fixed mass of gas trapped in a cylinder with a movable piston

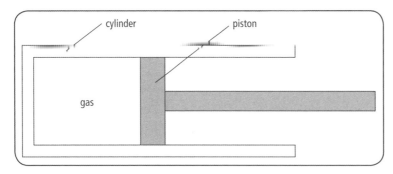

The temperature of the gas is increased.
  a. State what must happen to the piston, if anything, in order to keep the pressure of the gas constant.
  b. Explain your answer to part a.

CIE 0625 June '07 Paper 2 Q6b

3. a. State two changes that usually happen to the particles of a solid when the solid is heated.
   b. Most substances expand when they are heated.
      i)   State one example where such expansion is useful.
      ii) State one example where such expansion is a nuisance, and has to be allowed for.

CIE 0625 June '06 Paper 2 Q5

Thermal capacity

When heat energy is supplied to an object, its **internal energy** increases. This is because the molecules move faster so they have more kinetic energy. When an object's internal energy increases, the corresponding increase in temperature depends on its **thermal capacity**.

The thermal capacity of an object depends on the **material** from which it is made and its **mass**.

The **specific heat capacity** of a material is the energy required to raise the temperature of 1 kg of the material by 1 °C.

For example, The specific heat capacity of water = 4200 J/(kg °C), which means that it takes 4200 J of energy to raise the temperature of 1 kg water by 1 °C

Energy transferred = mass × specific heat capacity × temperature change

i.e.

$$E = mc\Delta T$$

Worked examples

1. Water has a specific heat capacity of 4200 J/(kg °C). How much heat energy must be supplied to raise the temperature of 1.5 kg of water by 10 °C?

2. 150 J of heat energy are required to raise the temperature of a 100 g block of metal by 5 °C. What is the specific heat capacity of the metal?

3. What is the increase in temperature of a 200 g block of metal of specific heat capacity 400 J/(kg °C) when 1500 J heat energy is supplied?

Answers

1. $E = mc\Delta T$
   $= 1.5 \times 4200 \times 10$
   $= 63\,000$ J
   $= 63$ kJ

2. $c = \dfrac{E}{m\Delta T}$
   $= \dfrac{150}{0.100 \times 5}$
   $= 300$ J/(kg °C)

3. $\Delta T = \dfrac{E}{mc}$

$ = \dfrac{1500}{0.200 \times 400}$

$ = 19\,°C$

Note: in questions 2 and 3 the mass in g must be changed into kg by dividing by 1000 i.e.

100 g = 0.100 kg

200 g = 0.200 kg

An experiment to determine specific heat capacity of water

Experimental arrangement

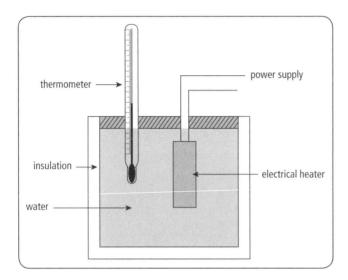

Procedure

Measure out 0.5 kg of water and pour into a beaker insulated with cotton wool. Place a thermometer in the water and cover it. Place an electrical heater of power 50 W in the water. Take the initial temperature of the water using the thermometer. Switch on the electrical heater and at the same time start a stop watch. Stir the water with the thermometer until the temperature has increased by 10 °C. Stop the stop watch and take the reading of the time taken for the water temperature to rise by 10 °C.

Processing the results

Time taken = 7 min = 420 s

Energy supplied by the heater = power × time = 50 × 420 = 21 000 J

Energy transferred to the water = $m \times c \times \Delta T$ = 0.5 × c × 10

Assuming all of the energy from the heater is transferred to the water,

0.5 × c × 10 = 21 000

$ c = 4200\ J/(kg\,°C)$

It is unlikely that a student would be able to obtain such an accurate value for c in an experiment, as **some heat is always lost to the surroundings**.

Melting

The following graph shows how the temperature of a substance varies with time as heat energy is supplied at a constant rate.

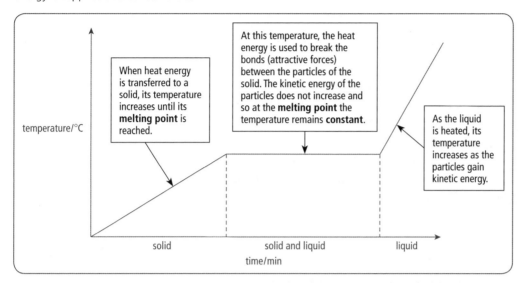

Solidification is the reverse of melting. A hot liquid loses heat to the surroundings which reduces its temperature. At the "melting point", the particles, instead of losing kinetic energy (causing a drop in temperature), arrange themselves into new "low energy" positions, i.e. the liquid becomes a solid.

The energy which must be put in to melt a solid at its melting point, or is given out when a liquid solidifies at its freezing point, is called the **latent heat of fusion**. There is no change in temperature.

Boiling

The following graph shows how the temperature of a substance varies as heat is supplied at a constant rate.

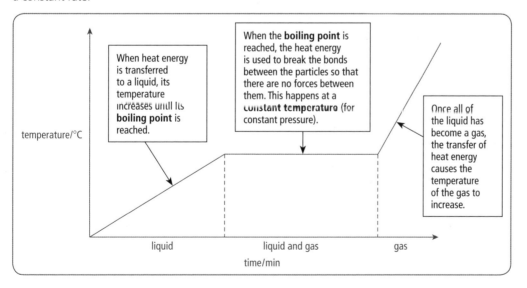

Boiling is different from **evaporation** in three ways:

- Boiling occurs at a fixed temperature, which depends on the substance being heated and its pressure.
- Evaporation can occur at all temperatures, including those below the boiling point.
- Evaporation decreases the temperature of the remaining liquid. The temperature of the liquid remains **constant** during boiling.

Condensation is the reverse of evaporation. If you breathe onto a cold window, you will see the water vapour (gas) in your breath turn back to liquid water as tiny droplets on the window. The window takes energy from the gas causing the particles to move to lower energy positions closer together, i.e. the gas becomes a liquid.

> The energy which must be put in to vaporise a liquid at its boiling point, or is given out when a gas at its boiling point condenses, is called the **latent heat of vaporisation**. There is no change in temperature.

The **specific latent heat of fusion** is the energy required to melt 1 kg of solid at its melting point, with no change in temperature.

The **specific latent heat of vaporisation** is the energy required to vaporise 1 kg of liquid at its boiling point, with no change in temperature.

For example, the specific latent heat of fusion for water = 300 000 J/kg which means that it takes 300 000 J of energy to melt 1 kg of pure ice at 0 °C.

> Energy transferred = mass × specific latent heat
>
> i.e.
>
> $$E = \Delta mL$$

Worked examples

1. What mass of water is changed from liquid to gas when 30 kJ of energy is supplied (assuming that there is no change in temperature)? The specific latent heat of vaporisation of water is 2 300 000 J/kg.

2. How much energy is required to melt 0.3 kg of ice? The specific latent heat of fusion of ice is 330 000 J/kg.

Answers

1. $\Delta m = \dfrac{E}{L}$

   $= \dfrac{30\ 000}{2\ 300\ 000}$

   $= 0.013$ kg

2. $E = \Delta mL$

   $= 0.3 \times 330\ 000$

   $= 99\ 000$ J

   $= 99$ kJ

Note: in question 1, kJ must be changed into J by multiplying by 1000

An experiment to find the latent heat of fusion of ice

Experimental arrangement

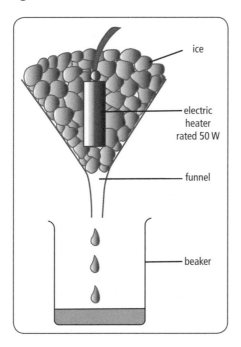

ice

electric
heater
rated 50 W

funnel

beaker

Procedure

Fill a funnel with ice and place a 50 W electric heater in the ice. Find the mass of the beaker using a balance. Place the beaker under the funnel and at the same time start a stop watch and switch on the electric heater. After 10 minutes, remove the beaker and switch off the heater. Find the mass of the beaker and water. Subtract the original mass of the beaker to find the mass of ice that has melted.

Processing the results

Energy supplied by the heater = power $\times$ time = $50 \times 600 = 30\,000$ J
Energy transferred to the water = $\Delta m \times L = 0.1 \times L$
Assuming only the energy from the heater is transferred to the melting ice,
$0.1 \times L = 30\,000$
$\qquad L = 300\,000$ J/kg

Inaccuracies in this experiment arise because some of the energy from the heater is lost to the surroundings and some energy from the surroundings is transferred to the ice. These two factors tend to cancel each other out and give a fairly accurate value for L.

An experiment to find the specific latent heat of vaporisation of water

Experimental arrangement

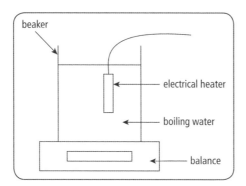

Procedure

Part fill a beaker with freshly boiled water and place on a balance. Place a 50 W electrical immersion heater in the water. Switch on the heater and wait for the water to boil. When the water is boiling, take the reading on the balance and at the same time start a stop watch. When the mass reading on the balance has decreased by 0.1 kg, take the reading on the stop watch.

Processing the results

Energy supplied by the heater = power × time = 50 × 4600
Energy transferred to the water = $\Delta m \times L$ = 0.1 × L
Assuming only the energy from the heater is transferred to the melting ice,
0.1 × L = 230 000
$\qquad L$ = 2 300 000 J/kg

Inaccuracies in this experiment arise because some of the energy from the heater is lost to the surroundings, which tends to give a value for L that is too large.

Examination style questions

1. An IGCSE student wishes to estimate the specific heat capacity of aluminium. She uses the experimental arrangement shown in the diagram.

    a. State the readings that she must take in order to calculate the specific heat capacity of aluminium.

    b. Suggest whether the calculated value of specific heat capacity will be higher or lower than the actual value.

    c. Explain your answer to part b.

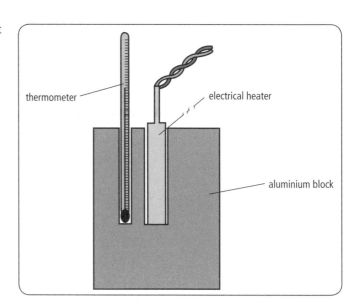

Adapted from CIE 0625 June '05 Paper 3 Q4

2. a. State two differences between evaporation and boiling.

b. Explain why energy is required to boil a liquid and why the temperature of the liquid remains constant during boiling.

c. A laboratory determination of the specific latent heat of vaporisation of water uses a 100 W heater to keep the water boiling at its boiling point. In 20 minutes, the mass of liquid water is reduced by 50 g. Calculate the value for the specific latent heat of vaporisation obtained from this experiment. Show your working.

Adapted from CIE 0625 June '06 Paper 3 Q4

3. Some water is heated electrically in a glass beaker in an experiment to find the specific heat capacity of water. The temperature of the water is taken at regular intervals.

The temperature-time graph for this heating is shown below.

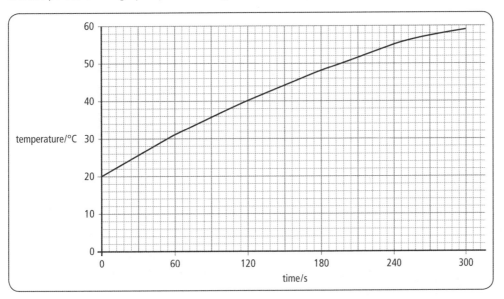

a. i) Use the graph to find
1. the temperature rise in the first 120 s,
2. the temperature rise in the second 120 s interval.
ii) Explain why these values are different.

b. The experiment is repeated in an insulated beaker. This time, the temperature of the water increases from 20 °C to 60 °C in 210 s. The beaker contains 75 g of water. The power of the heater is 60 W. Calculate the specific heat capacity of water.

c. In order to measure the temperature during the heating, a thermocouple is used. Draw a labelled diagram of a thermocouple connected to measure temperature.

CIE 0625 November '07 Paper 3 Q4

4. The diagram below shows apparatus that could be used to measure the specific latent heat of ice.

a. Describe how you would use the apparatus. You may assume that ice at 0 °C and a stopwatch are available. State all the readings that would be needed at each stage.

b. In an experiment, 120 g of ice at 0 °C is to be melted. The specific latent heat of ice is 340 J/g. Assume that all the energy from the heater will be used to melt the ice. Calculate the expected time for which the 60W heater is switched on.

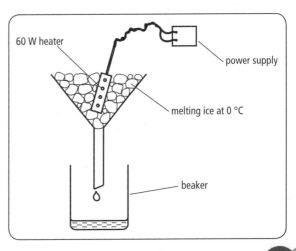

c. When the experiment is carried out, the ice melts in slightly less time than the expected time.
   i) State one reason why this happens.
   ii) Suggest one modification to the experiment that would reduce the difference between the experimental time and the expected time.

CIE 0625 November '05 Paper 3 Q4

Practical question

An IGCSE student is investigating the temperature rise of a beaker of water when heated by different methods. Beaker A is heated electrically and beaker B is heated by a Bunsen burner.

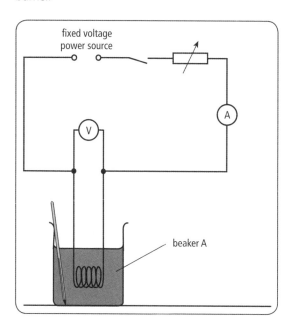
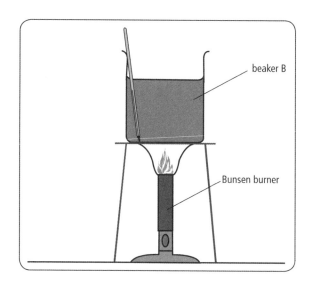

1. The student first records room temperature from the thermometer shown below.

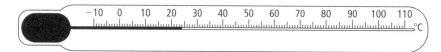

Write down the value of room temperature.

2. The beakers are both heated for 1 minute. Beaker A reaches a temperature of 32 °C and beaker B reaches a temperature of 29 °C. Calculate the temperature rise in each beaker.

3. The student expected the temperature rise in beaker A and beaker B to be the same. Give two possible reasons why the results were different.

2.3 Transfer of thermal energy

Conduction

An experiment to show that copper is a good conductor of heat

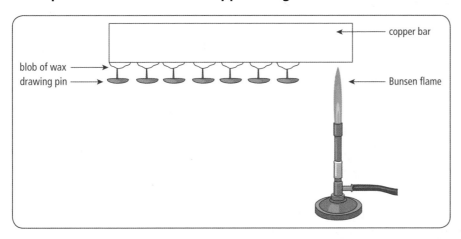

If a copper bar is heated at one end with a Bunsen flame, the drawing pins fall off one by one, beginning with the pin closest to the Bunsen flame. This is because as the metal **conducts** the heat from the hot end of the bar to the cold end, each of the blobs of wax melts in turn.

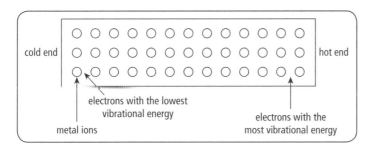

Metal atoms that have lost their free electrons are called **ions**. Sometimes metals are described as a lattice of metal ions in a "sea" of **free electrons**. Other solids are not usually good conductors of heat because they do not have free electrons to transfer heat energy from one molecule to another.

At the hot end of the metal bar, the ions gain energy and vibrate faster. The ions in a metal are close together and so these ions pass on their vibrations to neighbouring ions, which in turn start to vibrate faster. In this way, heat is transferred from the hot end to the cold end of the solid bar.

Metals also have free electrons in their structure which gain kinetic energy at the hot end of the bar. These free electrons pass on their kinetic energy through collisions with other electrons and metal atoms as they randomly diffuse through the metal. In this way, energy (heat) is conducted from the hot end of the bar to the cold end.

An experiment to show that water is a poor conductor of heat

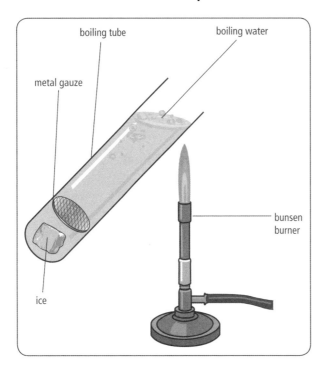

Ice is trapped at the bottom of the boiling tube with a piece of metal gauze. When the water at the top of the boiling tube is heated strongly, it boils. The ice at the bottom of the tube does not melt. This shows that water is a poor conductor of heat. However, if the ice is allowed to float normally, it melts quickly when the water is heated at the bottom of the test tube. This is because the water molecules can move, so the water heats by **convection**.

Water, like other liquids and non-metal solids, is a **poor conductor** of heat energy because its molecules do not have free electrons to easily pass on their kinetic energy to their neighbours, so the heat can only be transmitted through the vibration of the particles. Gases are **very poor conductors** of heat energy because their molecules are very far apart so kinetic energy cannot be transmitted from one molecule to another.

Materials that are poor conductors are called **insulators**. For example, air is an insulator.

Convection

An experiment to demonstrate convection in water

A few crystals of potassium permanganate are placed at the bottom of a beaker of water. They dissolve and colour the water near them purple. When the water is heated, the purple water rises above the Bunsen flame, moves across and then falls at the other side of the beaker before returning to the flame to be heated again.

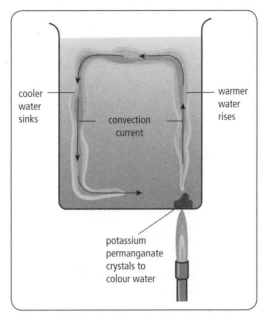

cooler water sinks

warmer water rises

convection current

potassium permanganate crystals to colour water

This movement of water is called a **convection current**.

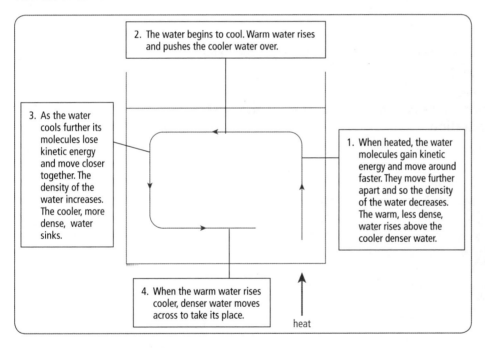

2. The water begins to cool. Warm water rises and pushes the cooler water over.

3. As the water cools further its molecules lose kinetic energy and move closer together. The density of the water increases. The cooler, more dense, water sinks.

1. When heated, the water molecules gain kinetic energy and move around faster. They move further apart and so the density of the water decreases. The warm, less dense, water rises above the cooler denser water.

4. When the warm water rises cooler, denser water moves across to take its place.

heat

Convection can only take place in fluids (liquids and gases) where the particles are free to move. Materials that have trapped air in them such as cotton wool or bubble wrap are good insulators, because air does not conduct heat, and trapped air cannot convect heat either.

Radiation

All hot objects emit infrared (thermal) radiation, part of the **electromagnetic spectrum** of waves (see section 3.2). Like all electromagnetic waves, infrared radiation can travel across a vacuum, which is why we are able to feel the heat of the Sun across the vacuum of space.

An experiment to demonstrate which type of surface is the best emitter of infrared radiation

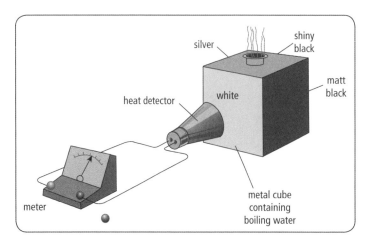

The metal cube has its vertical sides painted with four different surfaces: matt black, shiny black, white and silver. It is filled with boiling water and a heat detector (a thermopile) placed at a constant distance from it. The cube is rotated so that each of the sides faces the detector in turn and the reading on the meter noted.

The readings on the meter vary as follows:

| silver | white | shiny black | matt black |

Matt black surfaces are the best **emitters** of thermal radiation. Silver surfaces are the worst emitters of thermal radiation.

An experiment to demonstrate which colour surface is the best absorber of infrared radiation

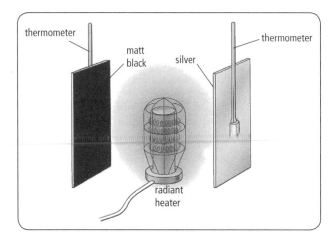

The initial readings are taken on the two thermometers. The radiant heater is switched on and the temperature on the two thermometers recorded at intervals of time.

Initially the readings on the thermometers are the same but after about 30 seconds the thermometer on the matt black surface gives a slightly higher reading than the thermometer on the silver surface. The temperature of the matt black surface continues to rise faster than the silver surface.

Matt black surfaces are the best **absorbers** of thermal radiation. Silver surfaces are the worst absorbers of heat radiation.

Examination style questions

1. The diagram below shows two containers, one painted matt black and the other shiny white. Both are filled with water, initially at the same temperature.

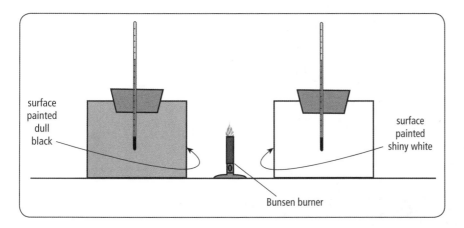

Describe how you would use the apparatus to determine which of the two surfaces is the better **absorber** of infrared radiation.

Adapted from CIE 0625 June '07 Paper 3 Q5a

2. Name the process by which thermal energy is transferred:
   a. from the Sun to the Earth.
   b. through the metal handle of a saucepan.

Adapted from CIE 0625 November '06 Paper 2 Q4a

Applications of conduction, convection and radiation

Vacuum flask

- The stopper is made of plastic, which is an insulator, to reduce heat flow by conduction. The stopper also stops heat flow by convection of air and stops heat flow by evaporation.
- The gap contains no air. So there are no particles to pass on the heat by conduction or convection.
- The silvered surfaces reflect infrared radiation and reduce heat flow by radiation.

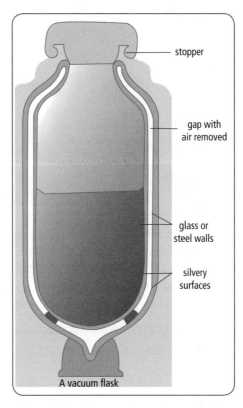

▲ A vacuum flask

Insulating the home

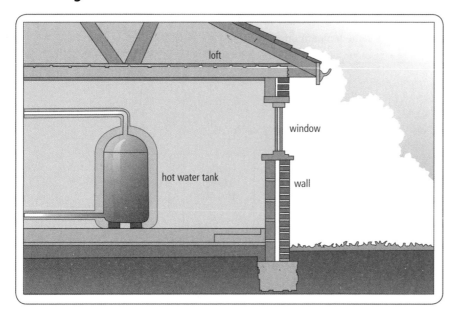

- "Lagging" around the hot water tank is made of plastic foam, which traps air to reduce heat loss by conduction (air is an insulator) and convection (the air is trapped and cannot move).
- Loft insulation is also made of fibres that trap air to reduce conduction and convection.
- An air cavity reduces heat exchange by conduction. Filling the air cavity with foam traps the air and also reduces convection.
- Double-glazed windows have air or another gas (argon or krypton) between the two layers of glass to reduce heat exchange by conduction and convection.

Sea breezes

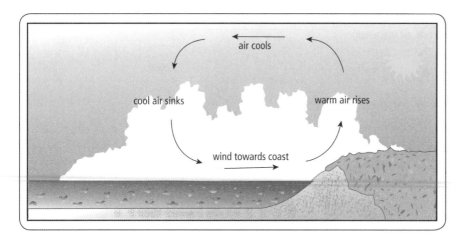

- During the day, **infrared radiation** from the Sun heats up the land more than the sea. The air above the land gets hotter than the air above the sea.
- The hot air above the land rises because it is less dense than the surrounding air.
- Cooler air from above the sea rushes in to take its place.
- This **convection current** causes a cool sea breeze to blow from the sea to the land.

Solar panel

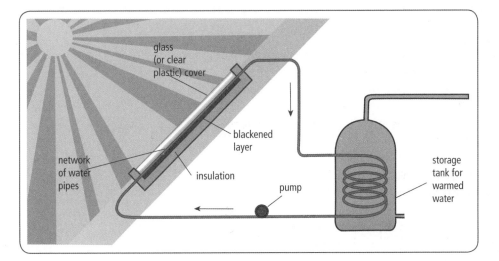

- The black surface of the solar panel absorbs **infrared radiation** from the Sun.
- The heat is then **conducted** through metal pipes to warm up water.
- The warm water rises to the top of the storage tank by **convection**.

Examination style questions

1. The diagram below shows a saucepan.

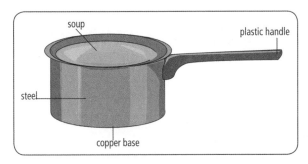

    a. Suggest why the base is made of copper.
    b. Suggest why the handle is made of plastic.
    c. The saucepan is placed on the hot hob of a cooker. The soup begins to heat up.
       i) Explain how the heat is transferred from the cooker hob to the soup.
       ii) By which of the following processes is heat transferred through the soup?
          conduction   radiation   convection   evaporation

CIE 0625 June '07 Paper 2 Q7a and b

2. An electric soldering iron is used to melt solder, for joining wires in an electric circuit.
A soldering iron is shown below.

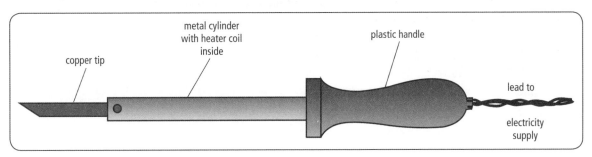

Solder is a metal which melts easily. The heater coil inside the metal cylinder heats the copper tip.

a. i)   Suggest why the tip is made of copper.
    ii) Suggest why the handle is made of plastic.

b. The heater coil is switched on. When the tip is put in contact with the solder, some of the heat is used to melt the solder.
    i)   State the process by which the heat is transferred from the copper tip to the solder.
    ii) By which process or processes is the rest of the heat transferred to the surroundings?
        conduction   convection   evaporation   radiation

c. A short time after switching on the soldering iron, it reaches a steady temperature, even though the heater coil is constantly generating heat.
    The soldering iron is rated at 40 W.
    What is the rate at which heat is being lost from the soldering iron?
        greater than 40 W
        equal to 40 W
        less than 40 W

CIE 0625 June '07 Paper 2 Q7

3. a. The diagram below shows a copper rod AB being heated at one end.

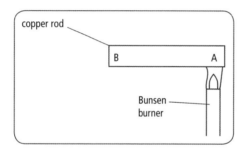

    i)   Name the process by which heat moves from A to B.
    ii) By reference to the behaviour of the particles of copper along AB, state how this process happens.

b. Give an account of an experiment that is designed to show which of four surfaces will absorb most heat radiation.

    The four surfaces are all the same metal, but one is a polished black surface, one is a polished silver surface, one is a dull black surface and the fourth one is painted white. Give your answer under the headings below.

    • labelled diagram of the apparatus
    • readings to be taken
    • one precaution to try to achieve a fair comparison between the various surfaces

CIE 0625 November '06 Paper 3 Q5

Practical question

The IGCSE class carries out an experiment to investigate the effect of insulation on the rate of cooling of hot water. The apparatus is shown in the diagram below.

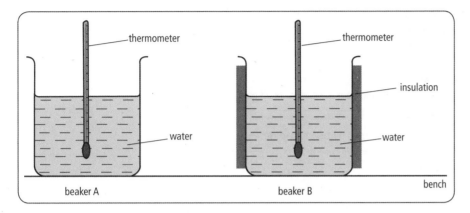

The students each have a stop watch, two glass beakers and some hot water. Beaker B is insulated.

A student fills beaker A about two thirds full with hot water and takes the initial reading on the thermometer, θ. At the same time the student starts a stop clock. He takes readings of the temperature every 30 seconds for four minutes. He then repeats the experiment for beaker B. The results of the experiment are recorded below.

Beaker A	
$t/$	$\theta/$
0	85
30	71
60	62
90	55
120	52
150	49
180	47
210	45
240	44

Beaker B	
$t/$	$\theta/$
0	85
30	74
60	65
90	57
120	55
150	51
180	49
210	47
240	46

1. What unit is missing from the column headings? Plot graphs of temperature (θ) against time (t) for beaker A and B on the same axes. Label both axes and draw the curves of **best fit**. Start the temperature axis at 40 °C.

2. The experiment was designed to investigate the effect of insulation on the cooling of hot water. Suggest two ways to improve the experiment.

Summary questions on unit 2

1. Match each heat transfer mechanism to its description

Conduction	An electromagnetic wave
Convection	Movement of hot fluid due to changes in density
Radiation	Most energetic particles escape from the surface of a liquid
Evaporation	Transfer of vibrational energy from particle to particle

2. Fill in the blanks

The specific latent heat of fusion is the energy required to _____ _____ kg

of solid. The specific latent heat of vaporisation is the energy required to _____

_____ kg of liquid. The specific heat capacity is the energy required to increase the

_____ of _____ kg of material by 1 °C.

3. Complete the table

	Solid	Liquid	Gas
Structure			
Arrangement of molecules		Close together but able to move past one another	
Compressible?			Yes
Flows?		Yes	
Shape	Keeps its shape		

4. The diagram below shows the coastline at night. Copy the diagram and draw four arrows to show the direction of the convection current in the air. Label each arrow to explain why the air is moving in the direction you have indicated.

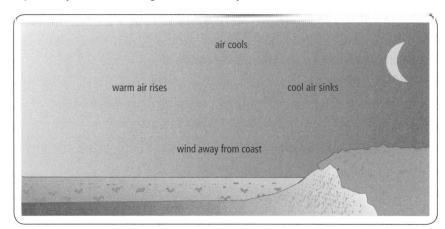

air cools

warm air rises   cool air sinks

wind away from coast

5. Write down the word equation used to calculate specific heat capacity.

Use the word equation to calculate the specific heat capacity of aluminium if 4500 J of energy is required to raise the temperature of 500 g by 10 °C. Show your working. Give the unit of specific heat capacity.

6. Write down the word equation used to calculate specific latent heat of fusion.

Use the word equation to calculate the specific latent heat of fusion of ice if 66 000 J of energy is required to melt 200 g of ice. Show your working. Give the unit of specific latent heat.

7. A sealed container with a moveable piston contains a fixed mass of dry air. The piston is used to reduce the volume of the gas without changing its temperature. Describe and explain the effect on the pressure of the gas

The piston is now fixed so that it cannot move and the volume cannot change. The gas is heated. Describe and explain the effect on the pressure of the gas.

8. Crossword

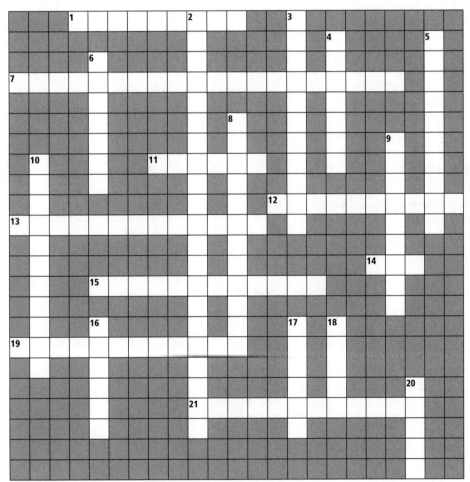

Across:
1 Heat transfer by electromagnetic wave (9)
7 The energy required to increase the temperature of 1 kg of substance by 1 °C (8, 4, 8)
11 Scale of temperature (6)
12 Heat transfer by passing on vibrations from one molecule to another (10)
13 Metals are good conductors because of these in their structure (4, 9)
14 A substance in which there are no forces between molecules (3)
15 Type of thermometer (12)
19 A solid changes into a liquid at this constant temperature (7, 5)
21 The temperature at which all molecular vibrations stop (8, 4)

Down:
2 The relationship between the pressure and volume of a fixed mass of gas at constant temperature (9, 12)
3 This happens in a puddle of water on a warm day (11)
4 Liquid often used in liquid-in-glass thermometer (7)
5 Heat transfer through the movement of hot fluid (10)
6 Energy of movement (7)
8 A liquid changes into a gas at this constant temperature (7, 5)
9 The colour of surface that is the best absorber of thermal radiation (4, 5)
10 An instrument for measuring temperature (11)
16 The colour of surface that is the worst emitter of thermal radiation (6)
17 Stops all heat transfer apart from radiation (6)
18 The unit of energy (5)
20 Substance in which the molecules vibrate in fixed positions (5)

9. Draw a mind map (spider diagram), including all the important points from this unit. Use diagrams, colour coding and mnemonics to help you remember the key points. Ensure that you group the key ideas logically. When you have finished, ask someone to test you on the content of your mind map.

Examination style questions on unit 2

1. The diagram below shows a student's attempt to estimate the specific latent heat of fusion of ice by adding ice at 0 °C to water at 20 °C. The water is stirred continuously as ice is slowly added until the temperature of the water is 0 °C and all the added ice has melted.

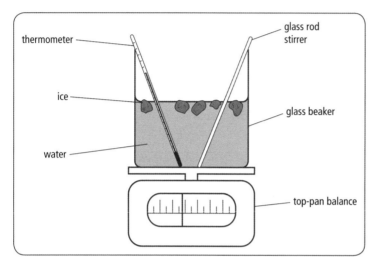

a. Three mass readings are taken. A description of the first reading is given. Write down descriptions of the other two.

   reading 1 the mass of the beaker + stirrer + thermometer

b. Write down word equations which the student could use to find
   i)   the heat lost by the water as it cools from 20 °C to 0 °C,
   ii) the heat gained by the melting ice.

c. The student calculates that the water loses 12 800 J and that the mass of ice melted is 30 g.
   Calculate a value for the specific latent heat of fusion of ice.

d. Suggest two reasons why this value is only an approximate value.

CIE 0625 June '07 Paper 3 Q4

2. The diagram below shows how the pressure of the gas sealed in a container varies during a period of time.

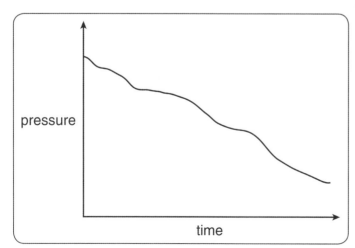

Which two of the following statements could explain this variation of pressure?

> The temperature of the gas is increasing.
> The temperature of the gas is decreasing.
> The volume of the container is increasing.
> The volume of the container is decreasing

CIE 0625 June '07 Paper 2 Q6

3. a. State two changes that usually happen to the molecules of a solid when the solid is heated.

   b. Most substances expand when they are heated.
      i)   State one example where such expansion is useful.
      ii) State one example where such expansion is a nuisance, and has to be allowed for.

CIE 0625 June '06 Paper 2 Q5

4. The diagram below shows a way of indicating the positions and direction of movement of some molecules in a gas at one instant.

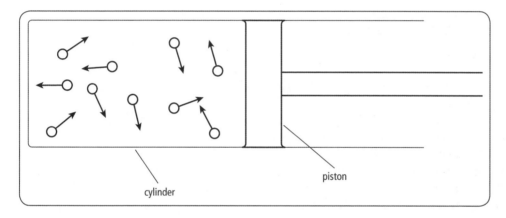

piston

cylinder

   a. i)   Describe the movement of the molecules.
      ii) Explain how the molecules exert a pressure on the container walls.
   b. When the gas in the cylinder is heated, it pushes the piston further out of the cylinder.
      State what happens to
      i)   the average spacing of the molecules,
      ii) the average speed of the molecules.
   c. The gas shown in the diagram is changed into a liquid and then into a solid by cooling.
      Compare the gaseous and solid states in terms of
      i)   the movement of the molecules,
      ii) the average separation of the molecules.

CIE 0625 November '05 Paper 3 Q5

3 Properties of waves

3.1 General wave properties

KEY IDEAS

✓ Every wave can be described in terms of its frequency, wavelength, velocity and amplitude
✓ In a longitudinal wave the vibrations of the particles are parallel to the direction of motion. In a transverse wave, the vibrations are perpendicular to the direction of motion
✓ All the particles along a wavefront are at the same point in their vibration
✓ Water waves, produced by a ripple tank, can be reflected, refracted and diffracted

A wave transfers energy from one place to another without transferring the particles of the medium. Individual particles vibrate (oscillate) about fixed positions.

There are two types of wave:

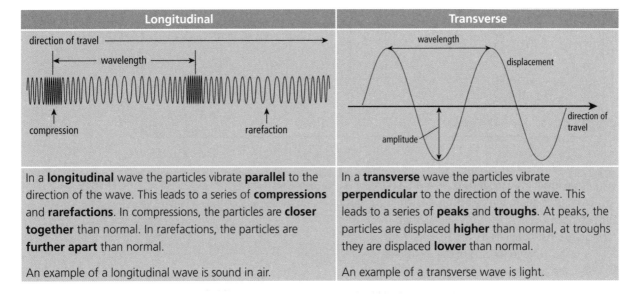

Longitudinal	Transverse
In a **longitudinal** wave the particles vibrate **parallel** to the direction of the wave. This leads to a series of **compressions** and **rarefactions**. In compressions, the particles are **closer together** than normal. In rarefactions, the particles are **further apart** than normal.	In a **transverse** wave the particles vibrate **perpendicular** to the direction of the wave. This leads to a series of **peaks** and **troughs**. At peaks, the particles are displaced **higher** than normal, at troughs they are displaced **lower** than normal.
An example of a longitudinal wave is sound in air.	An example of a transverse wave is light.

As shown on the two diagrams, the **wavelength is the distance between adjacent particles that are at the same point in their vibration** e.g. the distance between a compression and the next compression or the distance between a peak and the next peak.

The **amplitude** cannot easily be shown for the longitudinal wave but for the transverse wave it is the **distance from the centre of a vibration to the peak**, measured in m. The amplitude is the maximum displacement from the rest position.

The velocity of a wave is the distance travelled per second, measured in m/s.

The frequency of a wave is the number of complete waves passing a point every second, measured in hertz, Hz.

velocity = distance travelled by the wave in 1 second
frequency = number of waves passing a point every second or the number of
oscillations made by a particle on the wave every second.
The unit of frequency is **hertz (Hz)**

velocity (m/s) = frequency (Hz) × wavelength (m)

$$v \quad = \quad f \quad \times \quad \lambda$$

Worked examples

1. A sound wave has a frequency of 10 000 Hz and a wavelength of 0.033 m. What is the
speed of the wave?

2. A radio wave of speed 300 million m/s has a frequency of 600 MHz. What is its
wavelength?

3. What is the frequency of an ultrasound wave of speed 1500 m/s and
wavelength 0.05 m?

Answers

1. $v = f \times \lambda$
      = 10 000 × 0.033
      = 330 m/s

2. $\lambda = \frac{v}{f}$

   $= \frac{300\ 000\ 000}{600\ 000\ 000}$

   = 0.5 m

3. $f = \frac{v}{\lambda}$

   $= \frac{1500}{0.05}$

   = 30 000 Hz
   = 30 kHz

Note: 1 MHz = 1 million Hz = 1 000 000 Hz

 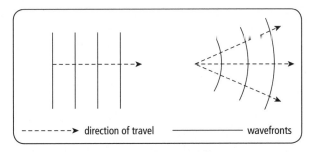

------> direction of travel ——————— wavefronts

Wavefronts can be represented as lines which are always **perpendicular** to
the **direction of wave travel**. The distance between one wavefront and the next is
one **wavelength**.

Ripple tank

The ripple tank has a vibrating bar attached to a motor, which can be used to set up waves of varying frequency in water.

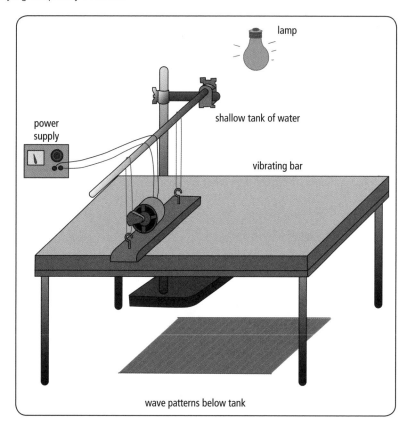

Water waves **reflect** at solid surfaces. There is no change in frequency, speed or wavelength on reflection.

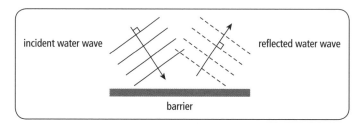

Wave theory suggests that each point on a wavefront can be considered to be a source of circular waves. These circular waves combine to make the wavefront and as they spread out, the wave travels forward.

Refraction

Water waves travel more slowly in shallow water.

As a wavefront AB approaches the boundary between the deep and shallow water, point A meets the shallower water first and this part of the wavefront **slows down**. Point B is still moving at the same speed as before. Wavefront AB turns clockwise which means that the new wavefront A'B' is now moving in a different direction. This change in direction is called **refraction**.

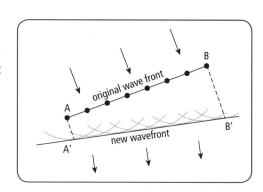

Summary questions on unit 3

1. Fill in the blanks

There are two type of wave: _____ and _____. In a _____

wave the oscillations are perpendicular to the direction of travel of the _____.

In a _____ wave the _____ are _____ to the direction

of travel. The _____ of a wave is the distance between adjacent points on the

wave that are at the same stage in their oscillation. The _____ _____

of a wave is the time for one complete wave. The frequency is the number of

_____ passing a point each _____. Velocity (or speed), frequency

and wavelength are related by the equation _____. When waves meet a

barrier, they bounce off. This is called _____. When waves speed up or

slow down the change direction. This is called _____. When waves pass

through a gap in a barrier, they spread out. This is called _____.

2. Mark amplitude and wavelength on the diagram below.

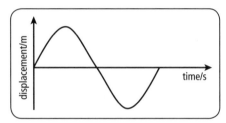

3. Complete the following diagrams to show the reflected and refracted rays.

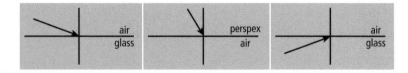

4. In each of the following cases, draw a diagram to show the incident, refracted and
reflected rays and the normal to the boundary. Calculate the angle of **refraction** and
state the angle of **reflection** in each case. The refractive index for air to glass is 1.5.
The refractive index for glass to air is 0.67.
   a.   Light incident at an angle of 56° to the normal in air on a glass surface.
   b.   Light incident in glass on an air/ glass boundary at an angle of 35° to the normal.
   c.   Light incident in glass at an angle of 60° to the normal at an air/glass boundary.

What is different about part c?

5. Match the terms on the right with the correct definition from the left hand column.

Ratio of speed of light in air to speed in material	Diffraction
Wave spreading out as it passes through a gap	Refractive index
Wave changing direction as its speed changes	Dispersion
Light bouncing off a smooth surface	Reflection
White light splitting into a spectrum as it is deviated by a prism.	Refraction

6. On graph paper, draw the following ray diagrams to scale:
   a. An object of height 2 cm at a distance of 5 cm from a convex lens of focal length 3 cm.
   b. An object of height 3 cm at a distance of 5.5 cm from a convex lens of focal length 2.5 cm.
   c. An object of height 2 cm at a distance of 2 cm from a convex lens of focal length 3 cm.

In each case, state a possible use for the arrangement.

7. a. What is the wavelength in air of electromagnetic radiation of frequency 1.0×10^{16} Hz? Which region of the spectrum does this belong to?
   b. What is the frequency of the electromagnetic radiation which has a wavelength of 0.1 m in air? Which region of the spectrum does this belong to?
   c. What is the wavelength in air of electromagnetic radiation of frequency 1.0×10^{20} Hz? Which region of the spectrum does this belong to?

8. Crossword

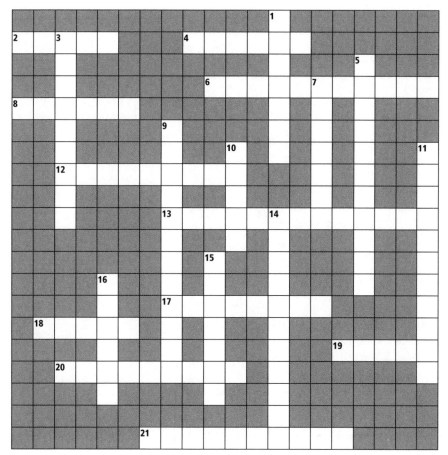

Across:
2 Region of the electromagnetic spectrum that has the highest energy (5)
4 The colour in the visible spectrum that is deviated the most by a prism (6)
6 Region between visible and X-rays in the electromagnetic spectrum (5-6)
8 Perpendicular to a boundary between two media (6)
12 The distance between the mid position of a particle on a wave and its maximum displacement. (9)
13 Angle of incidence when the angle of refraction is 90° (8, 5)
17 Region of the electromagnetic spectrum with longer wavelengths than visible light (5-3)
18 This type of wave cannot travel in a vacuum (5)
19 Region of the electromagnetic spectrum with the longest wavelength (5)
20 20 to 20 000 is the _____ range that the human ear can hear. (9)
21 Speed = frequency x _____ (10)

Down:
1 Water waves do this when they pass from deep to shallow water (7)
3 Electromagnetic radiation used to cook food quickly (9)
5 Distance between the centre of a lens and its principal focus (5,6)
7 An image that cannot be focused on a screen (7)
9 Carries information by total internal reflection of light (7, 5)
10 The unit of frequency (5)
11 Region of a sound wave where the pressure is lowest (11)
14 Region of a sound wave where the pressure is highest (11)
15 Sound waves do this when they hit a hard smooth surface (7)
16 Lens that is thicker in the middle than the edges (6)

9. Explain what is meant by dispersion.

10. Draw a diagram to show what happens to water waves, produced by a straight vibrating rod, when they are incident on a gap in a barrier.

11. A sound wave travels at 330 m/s in air with a wavelength of 0.1 m. What is its frequency? The wave enters water and the wavelength changes to 0.45 m. What is the speed of the sound wave in water?

12. A sound wave travels through a 2 m long metal rod in 0.42 ms. What is the speed of the wave?

Examination style questions on Unit 3

1. An inventor is trying to make a device to enable him to see objects behind him. He cuts
a square box in half diagonally and sticks two plane mirrors on the inside of the box.

A side view of the arrangement is shown below.

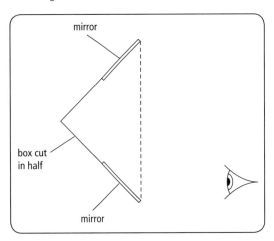

The following shows the arrangement, drawn larger. It shows parallel rays from two
different points on a distant object behind the man.

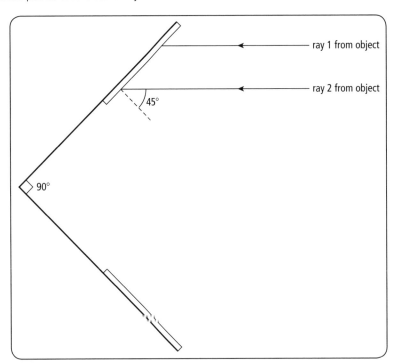

    a.   Carefully continue the two rays until they reach the place where the inventor's
head will be.

    b.   Look at what has happened to the two rays.
What can be said about the image the inventor sees?

CIE 0625 June '05 Paper 2 Q8

2. The speed of sound in air is 332 m/s. A man stands 249 m from a large flat wall, as shown below, and claps his hands once.

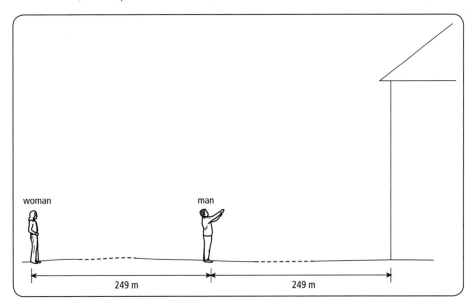

a. Calculate the interval between the time when the man claps his hands and the time when he hears the echo from the wall.

b. A woman is standing 249 m further away from the wall than the man. She hears the clap twice, once directly and once after reflection from the wall.
How long after the man claps does she hear these two sounds? Pick **two**.

0.75 s   1.50 s   2.25 s   3.00 s

CIE 0625 June '05 Paper 2 Q9

3. The diagram below shows the path of a sound wave from a source X.

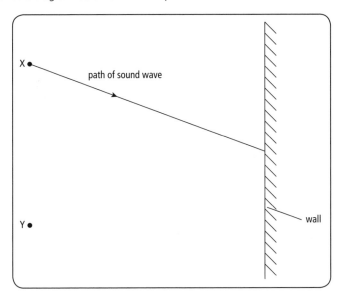

a. State why a person standing at point Y hears an echo.

b. The frequency of the sound wave leaving X is 400 Hz. State the frequency of the sound wave reaching Y.

c. The speed of the sound wave leaving X is 330 m/s. Calculate the wavelength of these sound waves.

d. Sound waves are longitudinal waves.
State what is meant by the term longitudinal.

CIE 0625 November '05 Paper 3 Q6

4. a. The diagram below shows two rays of light from a point O on an object. These rays are incident on a plane mirror.

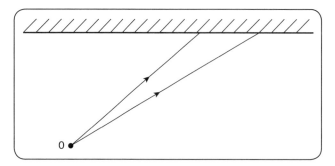

   i)   Copy the diagram and continue the paths of the two rays after they reach the mirror. Hence locate the image of the object O. Label the image I.
   ii) Describe the nature of the image I.

CIE 0625 November '05 Paper 3 Q7

5. Two students are asked to determine the speed of sound in air on the school playing fields.
   a. List the apparatus they need.
   b. List the readings that the students need to take.
   c. State how the speed of sound is calculated from the readings.
   d. State one precaution that could be taken to improve the accuracy of the value obtained.
   e. The table gives some speeds.

speed/m/s	speed of sound in air	speed of sound in water
10		
100		
1000		
10 000		

   Copy the table and place a tick to show the speed which is closest to
   i)   the speed of sound in air,
   ii) the speed of sound in water.

CIE 0625 June '07 Paper 3 Q7

6. The diagram below shows white light incident at P on a glass prism. Only the refracted red ray PQ is shown in the prism.
   a. Copy the diagram, draw rays to complete the path of the red ray and the whole path of the violet ray up to the point where they hit the screen. Label the violet ray.
   b. The angle of incidence of the white light is increased to 40°. The refractive index of the glass for the red light is 1.52. Calculate the angle of refraction at P for the red light.
   c. State the approximate speed of
   i)   the white light incident at P,
   ii) the red light after it leaves the prism at Q.

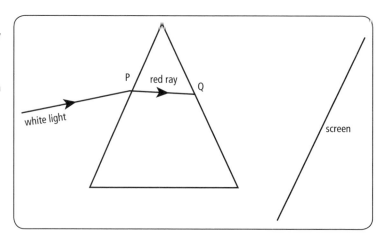

CIE 0625 June '06 Paper 3 Q6

4 Electricity and magnetism

4.1 Magnetism

KEY IDEAS

✓ Ferromagnetic materials can be magnetised and are attracted to magnets
✓ The field lines around a bar magnet can be plotted using compasses
✓ An electromagnet can be used to magnetise unmagnetised ferromagnetic objects

A north magnetic pole is actually a north seeking pole, i.e. if it is free to rotate it will point towards the Earth's north pole.

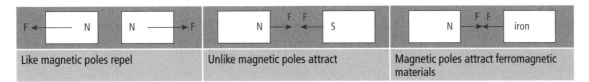

Like magnetic poles repel	Unlike magnetic poles attract	Magnetic poles attract ferromagnetic materials

Ferromagnetism

Iron is a ferromagnetic material. It contains what we can think of as tiny magnets called **domains**. In an unmagnetised piece of iron, the domains are arranged randomly but when a piece of iron is placed near a magnet, the domains line up because they are all attracted by the magnet.

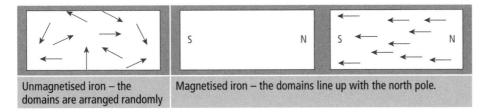

Unmagnetised iron – the domains are arranged randomly	Magnetised iron – the domains line up with the north pole.

This table shows how a magnet can pick up an unmagnetised piece of iron. When the domains line up, a south pole is formed opposite the north pole of the magnet. The poles attract. This is **induced** magnetism. When the magnet is removed, the piece of iron will quickly lose its magnetism. The domains become randomly arranged again and the **temporary magnetism** is lost.

Steel is made from iron and carbon and so it is also **ferromagnetic**. If steel is used instead of iron, the steel will keep become less strongly magnetised, but it will **retain** some of its induced magnetism and become a **permanent magnet**. It will remain magnetised until it is banged on the table or dropped.

Field around a bar magnet

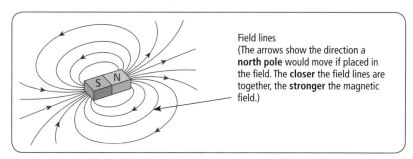

Field lines
(The arrows show the direction a **north pole** would move if placed in the field. The **closer** the field lines are together, the **stronger** the magnetic field.)

If small plotting compasses are placed around the bar magnet, the compasses show the direction of the magnetic field. If small pieces of iron (iron filings) are sprinkled around the magnet they too will line up with the magnetic field and show the field lines.

Electromagnetism

By convention, current is always shown as flowing from the positive side of a cell to the negative side in a complete circuit. **Conventional current is the flow of positive charge**.

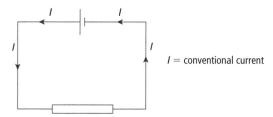

I = conventional current

Electrons are **negatively** charged, which means that they flow in the **opposite direction** to conventional current.

Current is measured using an **ammeter**, which must be placed in **series** in a circuit.

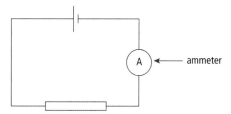

ammeter

When a current flows through a wire, a magnetic field is produced around the wire. The direction of the magnetic field depends on the direction of the current (from + to − around the circuit) and is given by the right hand grip rule as shown below.

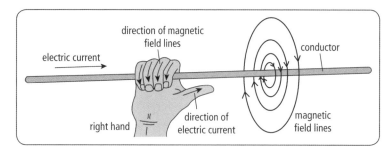

direction of magnetic field lines

electric current

conductor

right hand

direction of electric current

magnetic field lines

The **field lines** point in the direction of the **fingers** of your right hand when your **thumb** points in the direction of the **current**.

If the current flows in the opposite direction, then you must turn your hand round and see that the field lines now go round the wire in the opposite direction.

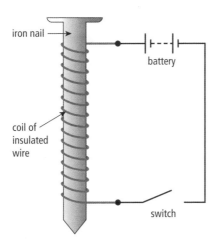

When a current flows through a coil of wire, a magnetic field is created around and inside the coil. The pattern of the **field lines** outside the coil is identical to a **bar magnet**. Objects made of ferromagnetic materials can be magnetised by placing them inside the coil.

An electromagnet can be switched off by opening the switch and cutting off the current from the battery. Its strength can be increased by increasing the voltage of the supply.

Relay

A relay is an electromagnetic switch that often uses a small current to switch on a much larger current, which may be dangerous if done directly.

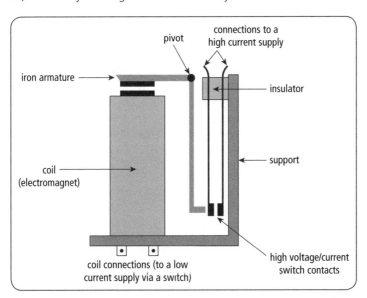

When the switch is closed, a small current flows through the coil, creating a magnetic field. The iron core of the coil becomes magnetised and attracts the iron armature, which pivots and pushes the contacts together. This closes the high current circuit and switches on heavy machinery, for example. The advantage of this arrangement is that the operator cannot come in contact with the high current supply.

Examination style questions

1. a. Explain what is meant by the **south pole** of a magnet.

b. The south poles of two magnets are brought together. Is the force between them attractive or repulsive or is there no force between them?

c. The iron bar is attracted to the north pole because of induced magnetism in the iron bar.

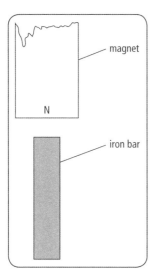

i) On the diagram, mark clearly the induced north pole and the induced south pole of the iron bar.

ii) State what happens to the induced magnetism in the iron bar when the magnet is taken away.

Adapted from CIE 0625 June '06 Paper 2 Q8

2. a. Two magnets are laid on a bench. End A of an unidentified rod is held in turn above one end of each magnet, with the results shown below.

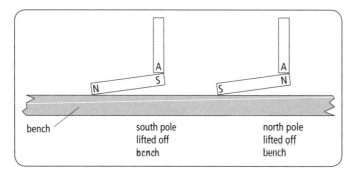

i) Suggest what the unidentified rod is made from.

ii) State what, if anything, happens when the end A is held over one end of
   1. an unmagnetised iron bar,
   2. an uncharged plastic rod.

b.   The diagram below shows four identical plotting compasses placed around a bar
     magnet where the magnetic field of the surroundings can be ignored. The pointer
     has only been drawn on one plotting compass.

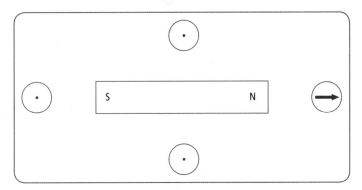

Copy the diagram, draw the pointers on the other three plotting compasses to
indicate the directions of the magnetic field of the bar magnet in those three places.

CIE 0625 November '05 Paper 2 Q8

3. The diagram below shows a bar magnet on a board in a region where the magnetic
        field of the surroundings is so weak it can be ignored. The letters N and S show the
        positions of the north and south poles of the magnet. Also on the diagram are marked
        four dots.

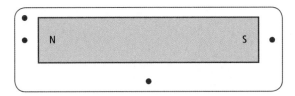

a.   Copy the diagram and carefully draw four magnetic field lines, one passing
     through each of the four dots. The lines you draw should begin and end either on
     the magnet or at the edge of the board.
b.   On one of your lines, put an arrow to show the direction of the magnetic field.

CIE 0625 June '07 Paper 2 Q11

4.2 Electrical quantities

KEY IDEAS

✓ The region around a charged particle, where another charge experiences a force, is called an electric field
✓ Current in amperes is the charge in coulombs passing a point every second
✓ Potential difference is the energy per coulomb of charge
✓ Resistance at constant temperature $= V/I$
✓ Power in a circuit $= VI$
✓ Energy transformed in a circuit $= VIt$

Electric charge

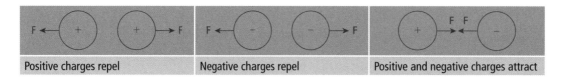

| Positive charges repel | Negative charges repel | Positive and negative charges attract |

The region around an electric charge where another charge experiences a force is called an electric field. The field lines show the direction a positive charge would move if placed in the field.

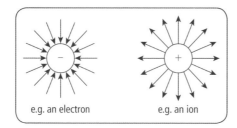

e.g. an electron   e.g. an ion

▲ Electric field lines for two source charges

Charging by induction

1.

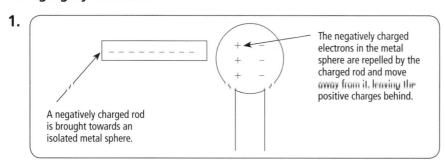

A negatively charged rod is brought towards an isolated metal sphere.

The negatively charged electrons in the metal sphere are repelled by the charged rod and move away from it, leaving the positive charges behind.

2.

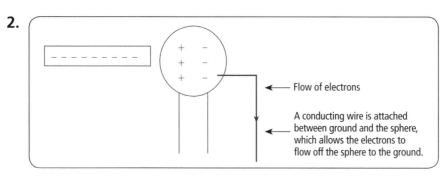

Flow of electrons

A conducting wire is attached between ground and the sphere, which allows the electrons to flow off the sphere to the ground.

110

3.

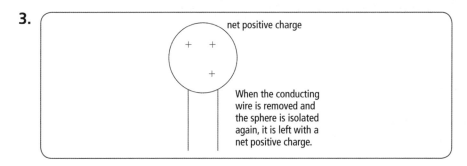

net positive charge

When the conducting wire is removed and the sphere is isolated again, it is left with a net positive charge.

Electric current

Charge is measured in **coulombs (C)**. Good **conductors**, such as metals, have **electrons** that are free to move through their structure. Poor conductors (**insulators**) do not have freely moving charged particles within their structure. Examples of insulators are rubber, plastic and paper.

The flow of charge through a conductor is called the **current**. Current will only flow through a conductor if there is a **potential difference** (PD) between the ends of the conductor.

> The current in amperes is equal to the charge in coulombs passing a point every second.
>
> Charge (C) = current (A) × time (s)
>
> $Q = \quad I \times t$

Worked examples

1. A charge of 0.01 C passes a point in a circuit every 0.2 s. What is the current flowing?

2. How long does it take for a charge of 30 C to pass a point in a circuit when a current of 0.8 A flows?

3. How much charge will pass a point in a circuit when a current of 0.5 A flows for 1 minute?

Answers

1. $I = \dfrac{Q}{t}$

$= \dfrac{0.01}{0.2}$

$= 0.05 \, \text{A}$

2. $t = \dfrac{Q}{I}$

$= \dfrac{30}{0.8}$

$= 37.5 \, \text{s}$

3. $Q = I \times t$

$= 0.5 \times 60$

$= 30 \, \text{C}$

Note: in question 3, 1 minute must be changed into 60 seconds.

Electromotive force and potential difference

An electrical supply (a power pack, cell or battery) provides electrical energy, which is carried round a circuit by the current. The electromotive force or **e.m.f.** of a supply is the energy per coulomb of charge, it is measured in **volts (V)**.

Potential difference or **voltage** across a component in a circuit is the energy required per coulomb of charge to drive the current through that component. It is measured in **volts (V)**.

The potential difference (**PD**) across a component is measured using a **voltmeter** in parallel with the component.

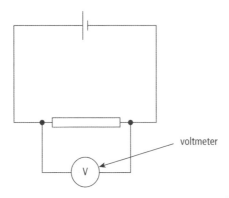

voltmeter

Resistance

Metals at a **constant temperature** have a constant resistance, measured in **ohms (Ω)**.

$$\text{Resistance } (\Omega) = \frac{\text{potential difference (V)}}{\text{current (A)}}$$
$$R = \frac{V}{I}$$

For a constant potential difference, **increasing the resistance decreases the current**.

For a constant resistance, **increasing the potential difference increases the current**.

Worked examples

1. A potential difference of 20 V is required for a current of 0.5 A to flow through a resistor. What is its resistance?

2. A current of 0.01 A flows through a resister of 1 kΩ. What is the potential difference across the resistor?

3. How much current flows when a potential difference of 5 V is applied to a resistor of 10 Ω?

Answers

1. $R = \frac{V}{I}$

$= \frac{20}{0.5}$

$= 40 \ \Omega$

2. $V = I \times R$

$= 0.01 \times 1000$

$= 10 \text{V}$

3. $I = \frac{V}{R}$

$= \frac{5}{10}$

$= 0.5 \text{ A}$

Note: in question 2 kΩ must be changed into Ω. 1 kΩ = 1000 Ω.

An experiment to find the resistance of an unknown resistor

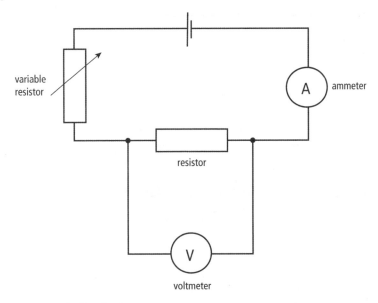

▲ Test circuit

Vary the potential difference across the unknown resistor by changing the resistance of the variable resistor. Measure the PD across the unknown resistor each time you change the resistance of the variable resistor, using the voltmeter and the corresponding values of current using the ammeter.

Results

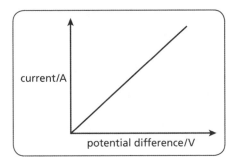

Analysis

$$\text{Resistance} = \frac{V}{I} = \frac{1}{\text{gradient of graph}}$$

Alternatively, resistance can be calculated by working out V/I for several values of V and I and an average resistance found.

The resistance of a wire

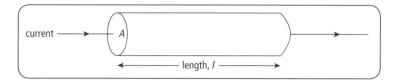

A metal wire has a length *l* and a cross-sectional area *A*. When the length of the wire is increased, the current has to travel further through the wire and so the resistance increases. When the cross-sectional area is increased by increasing the diameter of the wire, the current has a greater area to travel through and so the resistance decreases.

The resistance of the wire is directly proportional to the length of the wire and inversely proportional to the cross-sectional area.

Electrical power

Electrical energy is transferred from the battery or power supply in a circuit to the components in the circuit by the electrons. The component transforms the electrical energy into other forms (for example a bulb converts electrical energy into heat and light). The rate at which the energy is transformed is the **power**. Power can be calculated from the formula:

$$\text{Power (W)} = \text{potential difference (V)} \times \text{current (A)}$$

$$P = VI$$

$$\text{Energy} = \text{power} \times \text{time}$$

$$\text{so } E = VIt$$

Worked examples

1. What is the power of a bulb that allows a current of 1.5 A to flow when there is a potential difference of 8 V across it?

2. What is the current flowing through a bulb of power 40 W when there is a potential difference of 200 V across it?

3. What potential difference is required to produce a current of 0.3 A in a bulb of power 60 W?

Answers

1. $P = V \times I$

$= 8 \times 1.5$

$= 12\ \text{W}$

2. $I = \dfrac{P}{V}$

$= \dfrac{40}{200}$

$= 0.2\ \text{A}$

3. $V = \dfrac{P}{I}$

$= \dfrac{60}{0.3}$

$= 200\ \text{V}$

Examination style questions

1. The table below shows the potential difference (PD) needed at different times during a day to cause a current of 0.03 A in a particular thermistor.

a.   Calculate the two values missing from the table.

time of day (24-hour clock)	0900	1200	1500	1800
PD/V	15.0	9.9		7.5
resistance/Ω	500		210	250

b.   Copy the axes below, plot the four resistance values given in the table.

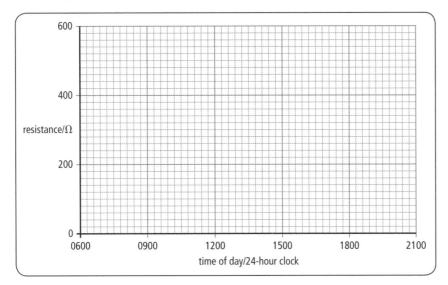

c.   i)   Draw a smooth curve through your points.
   ii)   Why do we draw a smooth curve rather than a series of straight lines joining the points?

d.   The thermistor is a circuit component with a resistance that decreases as the temperature increases.
   i)   From your graph, estimate the time of day when the temperature was greatest.
   ii)   State the reason for your answer to d(i).

CIE 0625 June '05 Paper 2 Q6

2. a.   i)   What name do we give to the type of material that allows electrical charges to pass through it?
   ii)   Give an example of such a material.
   iii)   What must be done to this type of material in order to make electrical charges pass through it?

b.   i)   What name do we give to the type of material that does **not** allow electrical charges to pass through it?
   ii)   Give an example of such a material.

c.   Which of the two types of material in a(i) and b(i) may be held in the hand and charged by friction (e.g. by rubbing with a soft cloth)?

CIE 0625 June '05 Paper 2 Q10

3. A student has a power supply, a resistor, a voltmeter, an ammeter and a variable resistor.

a. The student obtains five sets of readings from which he determines an average value for the resistance of the resistor.
    Draw a labelled diagram of a circuit that he could use.

b. Describe how the circuit should be used to obtain the five sets of readings.

c. The following circuit is set up and the reading on the ammeter is 0.5 A.

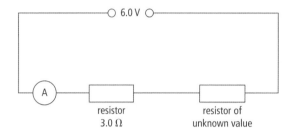

i)   Calculate the value of the unknown resistor.
ii) Calculate the charge passing through the 3.0 Ω resistor in 120 s.
iii) Calculate the power dissipated in the 3.0 Ω resistor.

CIE 0625 June '05 Paper 3 Q8

4. a. The diagram below shows a positively charged plastic rod, a metal plate resting on an insulator, and a lead connected to earth.

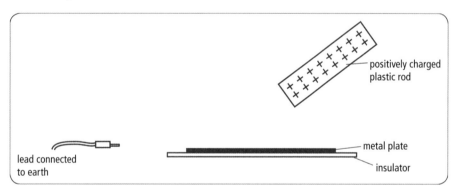

Describe how the metal plate may be charged by induction.

b. An electrostatic generator sets up a current of 20 mA in a circuit.
    Calculate
    i)   the charge flowing through the circuit in 15 s,
    ii) the potential difference across a 10 kΩ resistor in the circuit.

CIE 0625 June '06 Paper 3 Q10

Practical question

A student uses the circuit shown below to investigate the resistance of a piece of wire.

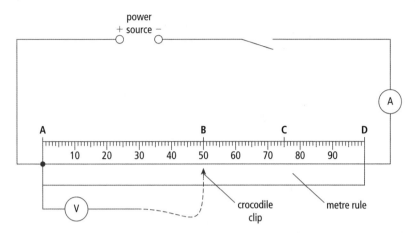

The student measures the current I in the wire. She then measures the PD V across **AB**, **AC** and **AD**.

The student's readings are shown in the table below.

section of wire	l/cm	I/A	V/V	R/
AB		0.375	0.95	
AC		0.375	1.50	
AD		0.375	1.95	

1. Using the diagram, record in the table the length l of each section of wire.

2. Show the positions of the pointers of the ammeter reading 0.375 A, and the voltmeter reading 1.50 V on the blank ammeter and voltmeter below.

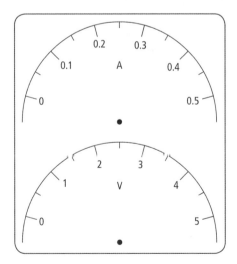

3. Calculate the resistance R of the sections of wire **AB**, **AC** and **AD** using the equation $R = V/I$.

4. Record these values of R, to a suitable number of significant figures, in the table.

5. Complete the column **heading** for the R column of the table.

6. Use your results to predict the resistance of a 1.50 m length of the same wire. Show your working.

4.3 Electric circuits

KEY IDEAS

✓ Current is the same in each part of a series circuit but is shared in a parallel circuit
✓ PD is the same across each part of a parallel circuit but is shared across a series circuit
✓ Thermistors and LDRs can be used as sensors
✓ Capacitors store charge and can be used to produce delays in circuits
✓ Digital signals have only two states: high and low
✓ Logic gates can be used to process signals to produce a desired output

Symbols for electrical components

Switch		Resistor	
Voltmeter	V	Ammeter	A
Lamp (1)		Lamp (2)	
Variable resistor		Cell	
Relay	NO COM NC	Bell	
Fuse		Transformer	
Diode		Transistor	
Thermistor		Light dependent resistor (LDR)	
Capacitor		Battery	

Drawing circuit diagrams

Use a pencil and a ruler to draw the connecting wires.
Do not draw the connecting wires through the components i.e. —(A)—
Always use the correct symbols for the components shown in the table above.

Series and parallel circuits

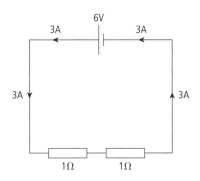

▲ Series circuit

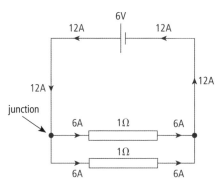

▲ Parallel circuit

118

In a series circuit the current is the **same** all the way round the circuit.	In a parallel circuit the current splits at the junction and is **shared** between the resistors.
Adding resistors in series **increases** the total resistance in the circuit.	Adding resistors in parallel **decreases** the total resistance of the circuit.
In a series circuit, the total resistance is simply the sum of the individual resistances. In the example above, **total resistance = 1 Ω + 1 Ω = 2 Ω**	In a parallel circuit, the total resistance is given by the formula $$\frac{1}{\text{total resistance}} = \frac{1}{R_1} + \frac{1}{R_2}$$ In the example above, $$\frac{1}{\text{total resistance}} = \frac{1}{1} + \frac{1}{1}$$ $$= 2$$ **total resistance $= \frac{1}{2} = 0.5\ \Omega$**
The potential difference across each resistor can be calculated using $V = IR$. In the example above, $V = 3 \times 1 = 3$ V, so each resistor has a PD of **3 V** across it.	The potential difference across each resistor can be calculated using $V = IR$. In the example above $V = 6 \times 1 = 6$ V, so each resistor has a PD of **6 V** across it.
In a series circuit, the potential difference is **shared** between the resistors. This is because the energy from the cell is shared between the resistors.	In a parallel circuit, the potential difference across each resistor is the **same** as the potential difference across the cell. This is because after picking up energy from the cell charge only passes through one of the resistors (not both).
If one of the resistors broke, the circuit would be broken and **no current** would flow.	If one of the resistors broke, the **current could still flow** through the second resistor, although the current would be smaller because there would now be a greater total resistance in the circuit.

Strings of party lights used to be wired in series, which meant if one of the bulbs broke, all of the bulbs would go out. Now, they are wired in parallel so that the rest of the bulbs will still work if one breaks.

Action and use of circuit components

Potentiometer

A potentiometer can be made from a **variable resistor**. Sliding the moving contact along the length of the resistor moves the contact represented by the arrow in the diagram. This changes the voltage on the voltmeter. Increasing the resistance in parallel with the voltmeter increases the share of the potential difference from the cell and the reading on the voltmeter increases.

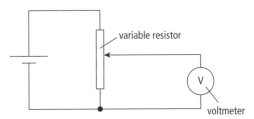

variable resistor

V

voltmeter

If the voltmeter is replaced with another component, the potential difference across the component can be varied with the potentiometer between 0 V and the maximum PD from the cell.

Transistor

A transistor is an electrically operated switch. It has three terminals: the **base**, **collector** and **emitter**. When a **small** current flows into the base, a **larger** current can flow between the collector and emitter. In this way the transistor also **amplifies the current**.

Thermistor

The thermistor is a component whose **resistance decreases as the temperature increases**. This means that it can be used as a **temperature sensor**.

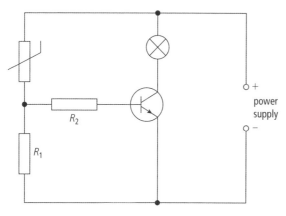

▲ Temperature sensitive circuit

When the thermistor is warmed, its resistance decreases and it takes a smaller share of the potential difference from the power supply. R_1 takes a larger share of the supply PD. The PD across the base is now large enough for the base current to switch on the collector–emitter current. When a large current flows from the collector to the emitter, the bulb lights.

This circuit could be used to turn on a warning light to show that the hob of an electric cooker is hot, or as an indicator light to show that hair straighteners have reached their operating temperature.

Light dependent resistor (LDR)

The light dependent resistor (LDR) is a component whose **resistance decreases as the light intensity increases**. This means that it can be used as a light sensor.

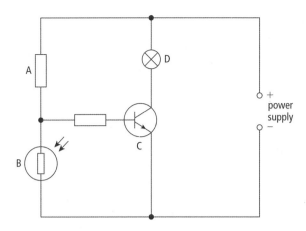

▲ Light sensitive circuit

When the light intensity on the LDR decreases, its resistance increases and it takes a larger share of the potential difference from the power supply. Resistor A takes a smaller share of the potential difference from the power supply. The PD across the base is now large enough for the base current to switch on the collector–emitter current. When a large current flows from the collector to the emitter, the bulb lights.

This circuit could be used to switch on a light at night.

Diode

A diode only allows current to flow **one way** through it, the direction in which the **arrow** is pointing.

This property of the diode is used in the conversion of a.c. current to d.c. Alternating current (a.c.) repeatedly changes the direction in which it flows around the circuit. Since the diode only allows current to flow in one direction, it changes the alternating current to direct current. This is called **rectification**.

Capacitor

A simple capacitor consists of two parallel metal plates separated by a layer of insulator. When a cell is connected across the capacitor plates, electric charge is forced onto the plates. If the cell is the disconnected, the capacitor plates remain charged until the capacitor is connected across a conductor. By storing charge, the capacitor is effectively storing **energy**.

When one capacitor plate is charged positive and the other negative, there is an electric field between the plates. In real capacitors, the plates and insulator are often rolled up together.

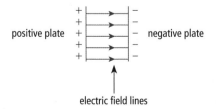

Capacitors can be used in a circuit where a component is to be switched on or off after a **time delay**.

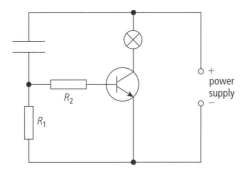

▲ Time delay circuit

As the capacitor charges, the potential difference across its plates increases. Initially R_1 gets a large share of the supply PD but this share decreases as the capacitor charges. When the PD across R_1 is low enough, the base current switches off the collector–emitter current. Initially, the bulb lights. After a time delay, which depends on the size of the capacitor and the resistance in the circuit, the bulb goes out.

This circuit could be used to switch on a hall light and then switch it off after a suitable time delay once the person had climbed the stairs.

Digital electronics

A digital system consists of an input sensor (such as a simple push switch) and a processor circuit, which controls the voltage to an output device (such as an electric bell). The processor circuit consists of a series of logic gates. Logic gates respond to small voltages, which are either on or off (digital signals). They do not respond to analogue signals.

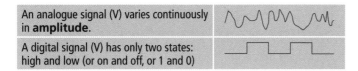

| An analogue signal (V) varies continuously in **amplitude**. | |
| A digital signal (V) has only two states: high and low (or on and off, or 1 and 0) | |

Logic gates
Logic gates are circuits containing transistors and other components. They transform a digital input voltage into an output, which depends on the type of logic gate. The input and output voltages are given as 1 or 0 (on or off) and can be represented in a truth table.

Logic gate	Symbol	Truth table
NOT	A ─▷o─ Y	A | Y 1 | 0 0 | 1
AND	A ─┐ ╲ B ─┘─ Y	A | B | Y 0 | 0 | 0 1 | 0 | 0 0 | 1 | 0 1 | 1 | 1
OR	A ─┐ ╲ B ─┘─ Y	A | B | Y 0 | 0 | 0 1 | 0 | 1 0 | 1 | 1 1 | 1 | 1
NAND	A ─┐ ╲o─ Y B ─┘	A | B | Y 0 | 0 | 1 1 | 0 | 1 0 | 1 | 1 1 | 1 | 0
NOR	A ─┐ ╲o─ Y B ─┘	A | B | Y 0 | 0 | 1 1 | 0 | 0 0 | 1 | 0 1 | 1 | 0

A NOT gate gives an output that is the opposite of the input.
An AND gate only gives an output if the input A *and* input B are both 1.
An OR gate gives an output if input A *or* input B is 1.

Logic gates can be combined to perform different functions in electrical circuits.

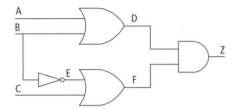

The truth table for the above arrangement is as follows:

A	B	C	D	E	F	Z
0	0	0	0	1	1	0
1	0	0	1	1	1	1
0	1	0	1	0	0	0
0	0	1	0	1	1	0
1	1	0	1	0	0	0
1	0	1	1	1	1	1
0	1	1	1	0	1	1
1	1	1	1	0	1	1

This is a very complicated logic gate circuit! It has three inputs A, B and C and a final output
Z. Start by reading columns A + B and looking at the output from the OR gate D. If either
A or B or both are high (1), then D is high (1). Now check the input B and the output E from
the NOT gate. The output from E is the opposite to the input at B. E + C are the inputs to
an OR gate whose output is F. Finally, look at the inputs D + F and the output Z to the AND
gate. Check that both D and F must be high (1) for Z to be high (1).

The following block diagram shows a circuit that will switch on a warning light at night
when the temperature is too low.

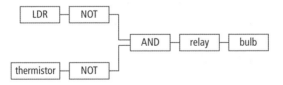

At night, the LDR output is 0, which is reversed by the NOT gate.
In the cold the thermistor output is 0, which is changed to 1 by the NOT gate.
This switches at the AND gate. The current is increased by the relay and the bulb lights.

Examination style questions

1.

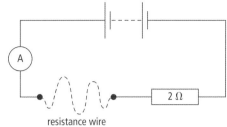

resistance wire

a. When the circuit above is connected up, how does the current in the resistance wire compare with the current in the 2 Ω resistor?

b. A voltmeter connected across the resistance wire shows the same reading as a voltmeter connected across the 2 Ω resistor. State the value of the resistance of the resistance wire.

c. Calculate the combined resistance of the wire and the resistor.

d. The wire and resistor are disconnected and then reconnected in parallel, as shown.

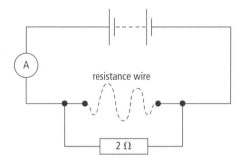

i) What is the combined resistance of the wire and resistor in the parallel circuit?

ii) The ammeter in the series circuit reads 0.3 A. Is the reading on the ammeter in the parallel circuit greater than, less than or equal to 0.3A?

e. Walls in buildings sometimes develop cracks. The width of a crack can be monitored by measuring the resistance of a thin wire stretched across the crack and firmly fixed on either side of the crack, as illustrated in the diagram below.

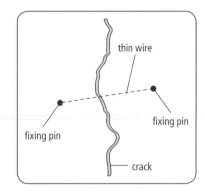

The wall moves and the crack widens slightly.
State what happens to
i) the length of the wire
ii) the resistance of the wire.

CIE 0625 June '05 Paper 2 Q11

2. The diagram below shows part of a low-voltage lighting circuit containing five identical lamps.

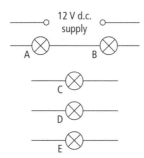

a. Complete the circuit, by the addition of components as necessary, so that
   i)   the total current from the supply can be measured.
   ii) the brightness of lamp E only can be varied.
   iii) lamps C and D may be switched on and off together whilst lamps A, B and E remain on.

b. All five lamps are marked 12 V, 36 W. Assume that the resistance of each lamp is the same fixed value regardless of how it is connected in the circuit.
   Calculate
   i)   the current in one lamp when operating at normal brightness.
   ii) the resistance of one lamp when operating at normal brightness.
   iii) the combined resistance of two lamps connected in parallel with the 12 V supply.
   iv) the energy used by one lamp in 30 s when operating at normal brightness.

c. The whole circuit is switched on. Explain why the brightness of lamps A and B is much less than that of one lamp operating at normal brightness.

CIE 0625 June '07 Paper 3 Q8

3. The circuit below is based on a transistor and thermistor

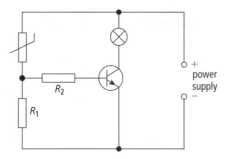

a. Describe the action of the thermistor in this circuit.
b. State and explain how the circuit may be modified so that the lamp switches on at a different temperature.
c. State one practical use of this circuit.

CIE 0625 June '07 Paper 3 Q10

4. The diagram below shows a low-voltage lighting circuit.

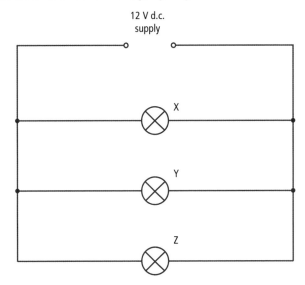

12 V d.c.
supply

a.   Copy the diagram, indicate with a dot and the letter S, a point in the circuit where a switch could be placed that would turn off lamps Y and Z at the same time but would leave lamp X still lit.

b.   i)   In the space below, draw the circuit symbol for a component that would vary the brightness of lamp X.

ii)   On your diagram, mark with a dot and the letter R where this component should be placed.

c.   Calculate the current in lamp Y.

d.   The current in lamp Z is 3.0 A. Calculate the resistance of this lamp.

e.   The lamp Y is removed.

i)   Why do lamps X and Z still work normally?

ii)   The current in lamp X is 1.0 A. Calculate the current supplied by the battery with lamp Y removed.

CIE 0625 November '06 Paper 3 Q8

5. The circuit below is used to switch on a lamp automatically when it starts to go dark.

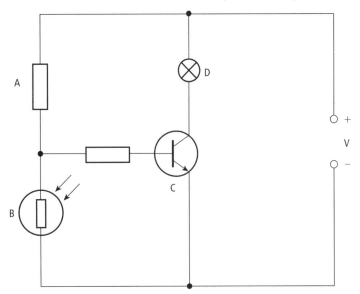

a.   Write down the names of the components labelled A, B, C and D.

b.   Which of the four components A, B, C or D acts as a switch?

c.   Explain why the lamp comes on as it goes dark.

CIE 0625 November '06 Paper 3 Q10

6. The diagram below shows a high-voltage supply connected across two metal plates.

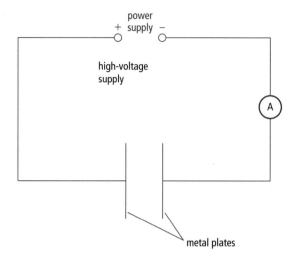

When the supply is switched on, an electric field is present between the plates.

a. Explain what is meant by an *electric field* .

b. Copy the diagram, draw the electric field lines between the plates and indicate their direction by arrows.

c. The metal plates are now joined by a high-resistance wire. A charge of 0.060 C passes along the wire in 30 s.
    Calculate the reading on the ammeter.

d. The potential difference of the supply is re-set to 1500 V and the ammeter reading changes to 0.0080 A. Calculate the energy supplied in 10 s. Show your working.

CIE 0625 November '05 Paper 3 Q8

7. a. Draw the symbol for a NOR gate. Label the inputs and the output.

b. State whether the output of a NOR gate will be high (ON) or low (OFF) when
    i)   one input is high and one input is low,
    ii) both inputs are high.

c. The diagram below shows a digital circuit made from three NOT gates and one NAND gate.

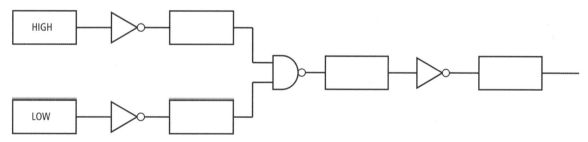

i)   Write HIGH or LOW in each of the boxes on the diagram
ii) State the effect on the output of changing both of the inputs.

CIE 0625 November '05 Paper 3 Q9

Practical question

A student is investigating the relationship between potential difference *V* across a resistor and the current *I* in it. The diagram below shows the apparatus that the student is using.

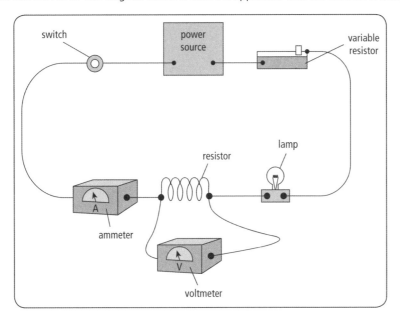

1. Draw the circuit diagram of the circuit shown above. Use standard circuit symbols.

2. The student is using a lamp to show when the current is switched on.
Why is it unnecessary to use the lamp?

3. State which piece of apparatus in the circuit is used to control the size of the current.

4. The student removes the lamp from the circuit. He is told that the resistance of a conductor is constant if the temperature of the conductor is constant. He knows that the current in the resistor has a heating effect. Suggest two ways in which the student could minimise the heating effect of the current in the resistor.

5. The diagram below shows a variable resistor with the sliding contact in two different positions.

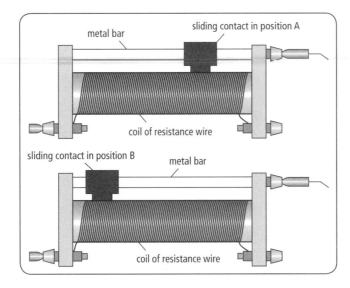

State which position, A or B, shows the higher resistance setting. Explain your answer.

4.4 Dangers of electricity

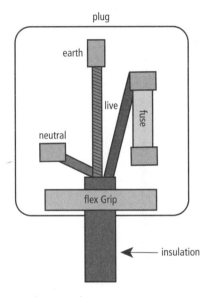

▲ The 3 pin plug

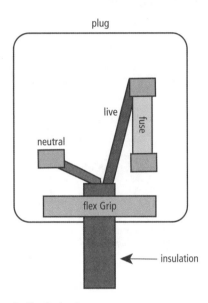

▲ The 2 pin plug

Safety features of the 3 pin plug and the 2 pin plug

• If the live wire comes loose and touches the metal casing of the appliance, the earth wire carries the current safely to ground and the fuse melts.
• In the 2 pin plug the earth wire is carried in grooves in the plastic case.
• If the current in an appliance becomes too large, say because there is a short circuit, the thin wire inside the fuse melts and breaks the circuit.
• The plastic coating over the cable insulates the conducting wires.

Water conducts electric current and so you should never touch an electrical appliance with wet hands or operate electrical equipment in wet conditions.

In the home, circuit breakers protect us from electrical fires.

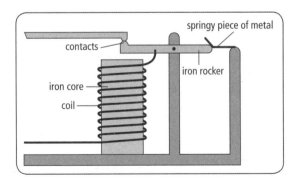

If the current flowing through the coil of wire becomes **too large**, the iron core becomes strongly magnetised and attracts the iron rocker with enough force to pull it down. This

opens the contacts and **breaks the circuit** so that the current can no longer flow. The circuit breaker can be reset by flicking a switch, which pushes down the springy piece of metal and pulls the rocker back up to close the contacts.

A "blown" fuse has to be replaced but a circuit breaker can easily be reset by lifting the iron rocker back into postion.

Examination style question

For each hazard, draw a line to the appropriate protection.

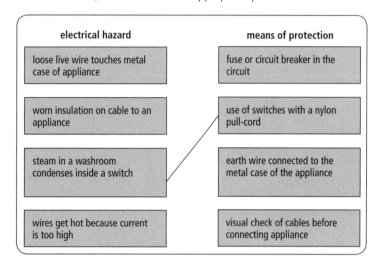

CIE 0625 June '06 Paper 2 Q12

4.5 Electromagnetic effects

KEY IDEAS

✓ When a coil cuts magnetic field lines, an e.m.f. is induced across the coil, this principle is used in the a.c. generator
✓ A transformer can be used to step up and step down a.c. voltage
✓ The direction of the force on a current carrying conductor in a magnetic field is given by the left hand rule
✓ A coil of wire carrying a current in a magnetic field will turn due to the forces on it; this principle is used in the d.c. motor.

Electromagnetic induction

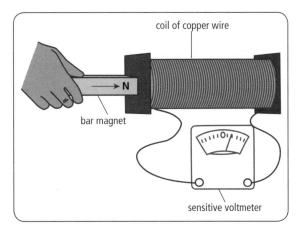

When the north pole of the bar magnet is moved into the coil, the needle on the sensitive voltmeter briefly moves to the right before returning to the centre.

The movement of the magnetic field lines due to the bar magnet **induces** an e.m.f. across the coil, which is then measured by the voltmeter. This is because as the magnet moves into the coil, the coil **cuts the magnetic field lines**.

When the north pole is moved out of the coil, the needle briefly moves to the left. This is because the coil is cutting the field lines in the opposite direction and so the **induced e.m.f.** is in the opposite direction.

The voltmeter can also be made to briefly move to the left if a south pole is moved into the coil.

Ways to increase the induced e.m.f.:

• Move the magnet faster
• Put more turns on the coil
• Use a stronger magnet

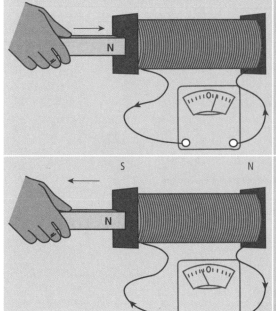

The e.m.f. across the coil can cause a current to flow in the coil if it is in a complete circuit. The current that flows causes the coil to act like a bar magnet. When a north pole is moved towards the coil, this induces a north pole at that end of the coil. **The e.m.f. opposes the change** because the north pole of the coil repels the north pole of the magnet.

When a north pole is moved away from the coil, this induces a south pole at that end of the coil. The south pole of the coil attracts the north pole of the magnet, again **opposing the change**.

a.c. generator

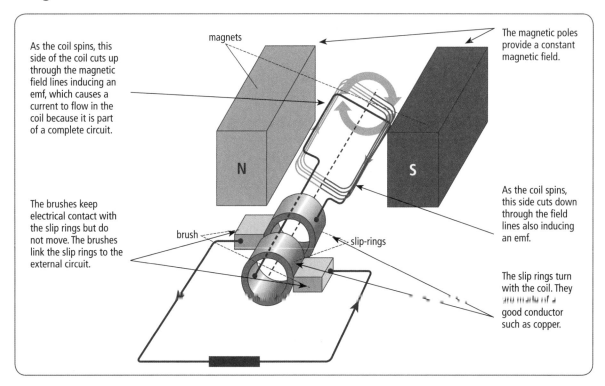

As the coil spins, this side of the coil cuts up through the magnetic field lines inducing an emf, which causes a current to flow in the coil because it is part of a complete circuit.

The brushes keep electrical contact with the slip rings but do not move. The brushes link the slip rings to the external circuit.

magnets

The magnetic poles provide a constant magnetic field.

As the coil spins, this side cuts down through the field lines also inducing an emf.

The slip rings turn with the coil. They are made of a good conductor such as copper.

brush · slip-rings

When the coil reaches the vertical position, the side that was previously cutting up through the field lines will now cut down through the field lines. This means that the **induced e.m.f.** will **change direction** and therefore the **current** will change direction.

Every time the coil reaches the vertical position, the current will change direction. The current produced is **alternating** or **a.c.**.

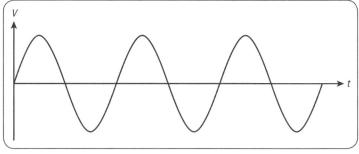

▲ Graph of voltage against time for an a.c. generator

Transformer

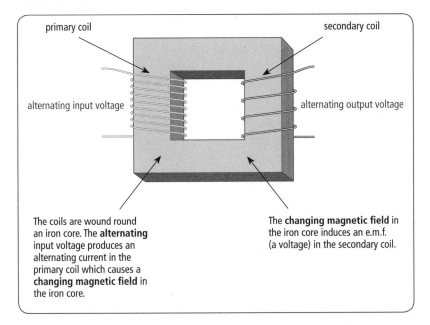

primary coil

secondary coil

alternating input voltage

alternating output voltage

The coils are wound round an iron core. The **alternating** input voltage produces an alternating current in the primary coil which causes a **changing magnetic field** in the iron core.

The **changing magnetic field** in the iron core induces an e.m.f. (a voltage) in the secondary coil.

This is a **step-down** transformer. The **output** voltage is **less** than the **input** voltage. This is because there are fewer turns on the primary coil than on the secondary.

A **step-up** transformer has **more** turns on the secondary coil than on the primary. The output voltage is **greater** than the input voltage.

$$\frac{V_s}{V_P} = \frac{N_s}{N_P}$$

V_s = voltage across the secondary coil; N_s = number of turns on the secondary coil (1 turn = 1 loop of wire in the coil);

V_P = voltage across the primary coil; N_P = number of turns on the primary coil

If the transformer is 100 % efficient, the input power is equal to the output power.
$P = IV$ so: $I_p V_p = I_s V_s$

Worked examples for you to try

1. The input voltage to a transform is 20 V. There are 100 turns on the primary coil and 6 000 turns on the secondary coil. What is the output voltage?

2. The input voltage to a transformer is 300 V and the output voltage is 12 V. There are 100 turns on the secondary coil. How many are these on the primary coil?

3. A transformer has 1 000 turns on the secondary coil and 200 turns on the primary coil. The output voltage is 10 V. What is the input voltage?

4. The input voltage to a transformer is 50 kV and the output voltage is 100 kV. There are 3 000 turns on the primary coil. How many turns are there on the secondary coil?

Answers

1. $\dfrac{Vs}{Vp} = \dfrac{Ns}{Np}$

$Vs = Vp \times \dfrac{Ns}{Np}$

$\quad = 20 \times \dfrac{600}{100}$

$Vs = 1200\ \text{V}$

i.e. There are 60 times as many turns on the secondary coil so there is 60 times as much voltage across this coil. Voltage and number of turns are directly proportional.

2. $\dfrac{Vs}{Vp} = \dfrac{Ns}{Np}$

$Np = Ns \times \dfrac{Vp}{Vs}$

$\quad = 100 \times \dfrac{300}{12}$

$Np = 2500$

i.e. There is 25 times as much voltage across the primary coil than the secondary and so there are 25 times as many turns on the primary coil

3. $\dfrac{Vs}{Vp} = \dfrac{Ns}{Np}$

$Vp = Vs \times \dfrac{Np}{Ns}$

$\quad = 10 \times \dfrac{200}{1000}$

$Vp = 2\ \text{V}$

i.e. There are 5 times fewer coils on the primary then on the secondary coil and so there is $\frac{1}{5}$ of the secondary voltage across the primary

4. $\dfrac{Vs}{Vp} = \dfrac{Ns}{Np}$

$Ns = Np \times \dfrac{Vs}{Vp}$

$\quad = 3000 \times \dfrac{100\,000}{50\,000}$

$Ns = 6000$

ie. the voltage is stepped up by a factor of 2 and so there are twice as many turns on the secondary coil than the primary.

Transformers are used to step up the voltage coming from a power station onto the power lines that transmit electrical energy. The power from the power station is constant and so **increasing the voltage decreases the current**. ($P = IV$ again) Transmitting electrical energy at high voltage and low current reduces the energy lost as heat from the power lines and **increases the efficiency** of the system.

Force on a current–carrying conductor

A wire carrying a current has a magnetic field around it (direction given by the right hand grip rule). If the wire is placed in another magnetic field, the two magnetic fields will interact and there will be a **force** on the wire.

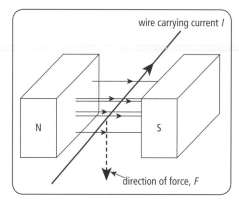

In the above arrangement, the wire will move **downwards** due to the force on it.

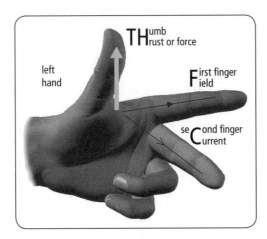

The direction of the force is given by the **Fleming's left hand rule**.

To check the direction of the force, hold your fingers in the position shown in the picture. Now turn your hand so that your second finger points into the page and your first finger points from left to right across the page. Your thumb should point down the page, in the direction of the force.

Remember: **F**irst finger **F**ield, se**C**ond finger **C**urrent, thu**M**b for **M**otion

• If the direction of the magnetic field stays the same but the direction of the current is reversed (so that it comes out of the page), the force reverses and is now upwards.
• If the direction of the current remains into the page and the magnetic field is reversed (so that it goes right to left across the page), the force reverses and is now upwards.

Check this example using the left hand rule.

The d.c. motor

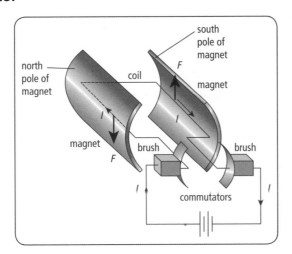

▲ The d.c. motor

• The poles of the **magnet** are curved to provide a radial magnetic field. This helps to keep the coil in a constant magnetic field.
• There is a force, **F**, on each side of the coil since the coil is carrying a current in a magnetic field. The direction of these forces is given by the left hand rule. The forces cause the coil to **spin**.

- The **commutator** turns as the coil turns. It keeps in electrical contact with the **brushes** and so the current, **I**, keeps flowing through the coil. Every time the coil reaches the vertical position, the two sides of the commutator swap brushes and the **flow of current is reversed**. This means the direction of the **force** on each side of the coil is **reversed** and so the coil keeps spinning in the same direction. Without the commutator, the coil would just oscillate backwards and forwards.

The speed at which the coil spins can be increased by

- increasing the number of turns on the coil
- increasing the current
- increasing the strength of the magnetic field

Examination style questions

1. A person has a 6 V bell. He hopes to operate the bell from a 240 V a.c. mains supply, with the help of the transformer shown below.

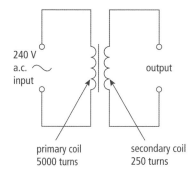

240 V
a.c.
input

output

primary coil
5000 turns

secondary coil
250 turns

a. State how you can tell from the diagram that the transformer is a step-down transformer.
b. State how the output voltage compares with the input voltage in a step-down transformer.
c. Calculate the output voltage of the transformer when connected to the 240 V mains supply.
d. Why would it not be wise for the person to connect the 6 V bell to this output?

CIE 0625 November '06 Paper 2 Q10

2. The diagram below shows a flexible wire hanging between two magnetic poles. The flexible wire is connected to a 12 V d.c. supply that is switched off.

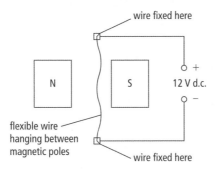

a.   Explain why the wire moves when the supply is switched on.

b.   State the direction of the deflection of the wire.

c.   When the wire first moves, energy transfers from one form to another. State these two forms of energy using an arrow to show the direction of transfer.

d.   The diagram below shows the flexible wire made into a rigid rectangular coil and mounted on an axle.

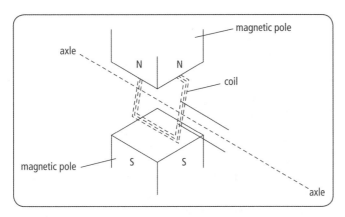

i)   Add to the diagram an arrangement that will allow current to be fed into the coil whilst allowing the coil to turn continuously. Label the parts you have added.

ii)   Briefly explain how your arrangement works.

CIE 0625 June '05 Paper 3 Q11

3. A transformer is needed to step down a 240 V a.c. supply to a 12 V a.c. output.

a.   Draw a labelled diagram of a suitable transformer.

b.   Explain

i)   why the transformer only works on a.c.,

ii)   how the input voltage is changed to an output voltage.

c.   The output current is 1.5 A. Calculate

i)   the power output

ii)   the energy output in 30 s.

CIE 0625 June '06 Paper 3 Q9

4. The diagram below shows the basic parts of a transformer.

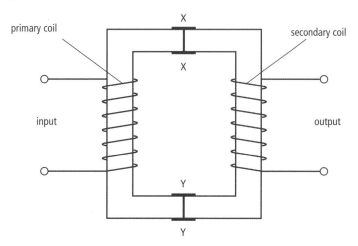

primary coil   X   secondary coil

X

input   output

Y

Y

   a.   Use ideas of electromagnetic induction to explain how the input voltage is
        transformed into an output voltage. Use the three questions below to help you
        with your answer.
        What happens in the primary coil?
        What happens in the core?
        What happens in the secondary coil?
   b.   State what is needed to make the output voltage higher than the input voltage.
   c.   The core of this transformer splits along XX and YY. Explain why the transformer
        wouldnot work if the two halves of the core were separated by about 30cm
   d.   A 100% efficient transformer is used to step up the voltage of a supply from 100V to
        200V. A resistor is connected to the output. The current in the primary coil is 0.4A.
        Calculate the current in the secondary coil.

CIE 0625 November '05 Paper 3 Q10

5. Your teacher gives you a length of wire, a sensitive millivoltmeter and a powerful
    magnet. You are asked to demonstrate the induction of an e.m.f. in the wire.
   a.   Describe what you would do.
   b.   How would you know that an e.m.f. has been induced?
   c.   Name a device which makes use of electromagnetic induction.

CIE 0625 June '07 Paper 2 Q10

Cathode ray tube

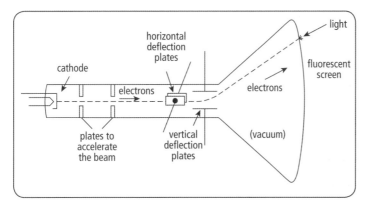

- The "electron gun" consists of a filament (the **cathode**) that produces electrons when heated by **thermionic emission**.
- The electrons are accelerated through a potential difference, **away from the negatively charged cathode**.
- The horizontal and vertical deflection plates **change the direction** of the beam of electrons, using an electric field. The electrons are attracted by the positively charged plate and repelled by the negative plate.
- When the beam of electrons hits the **fluorescent** screen, their kinetic energy is transformed to **light** energy and a bright spot appears.

The cathode ray oscilloscope (CRO) can be used as a high resistance voltmeter. When it is connected in parallel with a component in a circuit a bright trace appears on the screen. The trace shows how the **amplitude** (size) of the **PD** varies with **time**.

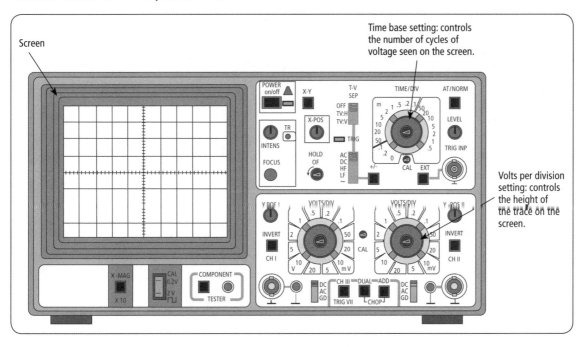

▲ Typical cathode ray oscilloscope

Examples of traces on the oscilloscope screen

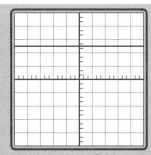

Y gain setting = 2 V per division

This is the trace obtained for a **d.c.** voltage of **5 V** (2.5 divisions × 2 V = 5 V)

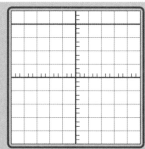

Y gain setting = 10 mV per division

This is the trace obtained for a **d.c.** voltage of **40 mV** (4 divisions × 10 mV = 40 mV)

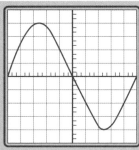

Y gain setting = 0.5 V per division
Time base setting = 1 s per division

This is the trace obtained for an **a.c.** voltage with a **peak voltage of 2 V** (4 divisions × 0.5 V per division) and a **peak to peak voltage of 4 V.**

The time period = time for one complete cycle of voltage = **10 s** (10 divisions × 1 s per division).

Frequency = $\dfrac{1}{\text{time period}}$ = **0.1 Hz**

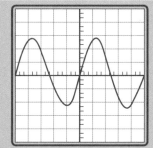

Y gain setting = 1 V per division
Time base setting = 10 ms per division

This is the trace obtained for an **a.c.** voltage with a **peak voltage** of **2.8 V** (2.8 divisions × 1 V per division) and a **peak to peak voltage** of **5.6 V.**

The time period = time for one complete cycle of voltage = **50 ms** (5 divisions × 10ms per division).

Frequency = $\dfrac{1}{\text{time period}}$ = $\dfrac{1}{0.05}$ = **20 Hz**

Examination style questions

1. a.   The diagram below shows how a beam of electrons would be deflected by an electric field produced between two metal plates. The connections of the source of high potential difference are not shown.

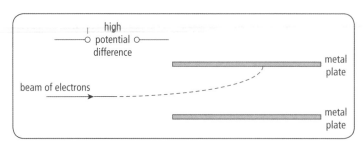

   i)   Draw in the missing connections. Complete the diagram by inserting the high potential difference in the correct place.

   ii) Explain why the beam of electrons is deflected in the direction shown. In your answer, consider all the charges involved and their effect on each other.

b. The deflection of a beam of electrons by an electric field is used in cathode ray oscilloscopes.

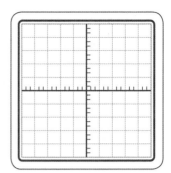

   i)   What makes the electron beam move backwards and forwards across the screen?
   ii) What makes the electron beam move up and down the screen?

c. An a.c. waveform is displayed so that two full waves appear on the screen of a cathode ray oscilloscope.

d. The diagram on the right shows the face of the oscilloscope. On the diagram, draw in the waveform that will be displayed.

Adapted from CIE 0625 November '06 Paper 3 Q9

2. a. The diagram below shows a type of tube in which cathode rays can be produced.

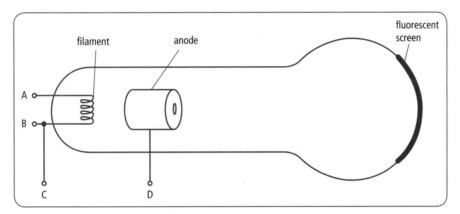

   i)   A PD is connected between two terminals in order to cause thermionic emission. Between which two of the four labelled terminals is the PD connected?
   ii)   Where does the thermionic emission occur?
   iii) Which particles are emitted during thermionic emission?
         α-particles
         electrons
         neutrons
         protons

   iv)   Draw the path of the cathode rays that are created when all the electrical connections are correctly made.
   v)   State what is seen when the cathode rays strike the fluorescent screen.

b. The diagram below shows the same tube as previously, with two metal plates alongside the tube. A high PD is connected between the plates.

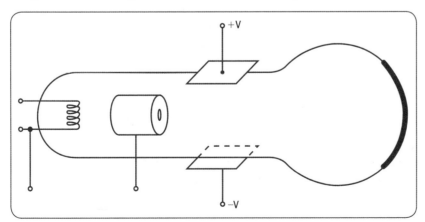

   Draw the path of the cathode rays.

c. The tube in both diagrams has a vacuum inside it.
   State why this vacuum is necessary.

Summary questions on unit 4

1. What are the units for the following quantities?

Charge	Time period
Current	Frequency
Potential difference or voltage	Energy
Resistance	Power

2. Complete the table to show the names of the electrical components and their symbols

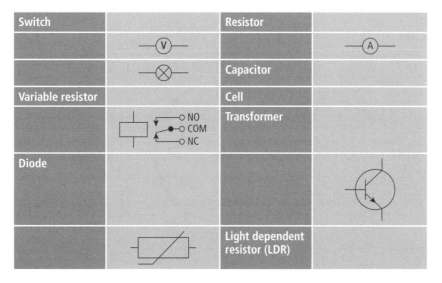

Switch		Resistor	
	—(V)—		—(A)—
	—⊗—	Capacitor	
Variable resistor		Cell	
	NO COM NC	Transformer	
Diode			
		Light dependent resistor (LDR)	

3. Write down the equation linking current, charge and time. Use the correct symbols. Then answer the following questions. Give the answer with the correct unit.
  a. A current of 3 A flows for 6 s. How much charge passes a point in that time?
  b. 6 mC of charge pass a point in a circuit every 100 s. What is the current in the circuit?
  c. How long does it take for 15 C of charge to pass a point when a current of 0.1A flows?

4. Write down the equation linking resistance, potential difference and current. Use the correct symbols. Then answer the following questions. Give the answer with the correct unit.
  a. What is the potential difference across a 100 Ω resistor when a current of 0.5 A flows through it?
  b. What current flows through a resistor of 330 Ω when a there is a potential difference of 10V across it?
  c. What value of resistor has a current of 0.8 A flowing through it when the potential difference across it is 1.5 V?

5. Write down the equation linking current, potential difference and power. Use the correct symbols. Then answer the following questions. Give the answer with the correct unit.
  a. What is the power of a lamp when a current of 3 A flow through it and the potential difference across it is 12 V?
  b. A lamp of power 40 W has a potential difference of 230 V across it. What is the current through the lamp?
  c. A lamp of power 100 W has a current of 0.5 A flowing through it. What is the potential difference across the lamp?

6. Explain how an earth wire and a fuse can protect a person from an electric shock.

7. Draw a circuit to vary the potential difference across a lamp and measure the current through the lamp. Include an appropriate component to measure the potential difference across the lamp and a switch to switch the circuit on and off.

8. Draw the symbols for a NOT, AND and OR gate and give their truth tables.

9. Explain how a relay can be used to switch on a large voltage using a small voltage.

10. Draw the electric field lines between the plates of a capacitor.

11. Complete the paragraph

When a current carrying conductor is placed in a magnetic _____, there is
a _____ on the conductor. The direction of the _____ is given by
the _____ _____ rule, where the first finger is the _____,
second finger is the _____ and the thumb is for _____. The force
on a conductor in a magnetic field can be increased by increasing the _____
through the conductor or increasing the _____ of the _____
_____.

12. Draw and label a diagram of a d.c. motor. Explain how the commutator allows the motor to turn continuously.

13. Describe an experiment to show that the direction of the e.m.f. induced across a conductor depends on the direction that the conductor is moved through a magnetic field.

14. Draw and label a diagram of an a.c. generator. Explain how the slip rings enable the generator to turn continuously.

15. Give three ways in which the output voltage from an a.c. generator can be increased.

16. Draw a diagram of a step up transformer. Label the primary and secondary coils and the iron core. Explain why the transformer will not operate with a d.c. input voltage.

17. Use the information given to sketch an oscilloscope trace for the following examples.

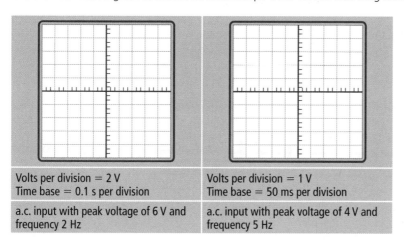

Volts per division = 2 V Time base = 0.1 s per division	Volts per division = 1 V Time base = 50 ms per division
a.c. input with peak voltage of 6 V and frequency 2 Hz	a.c. input with peak voltage of 4 V and frequency 5 Hz

18. Crossword

Across:

1 A component whose resistance decreases with increasing temperature (10)
3 Equal to current x voltage in an electrical circuit (5)
6 An electrical component that stores charge (9)
7 Used to determine the direction of the force on a current carrying conductor in a magnetic field (4, 4, 4)
8 Limits electrical current in wiring in the home (7, 7)
11 A property exhibited by the elements iron, cobalt and nickel (14)
12 Produced when a conductor cuts magnetic field lines (7, 3)
15 Repels positive charges (8, 6)
16 Used to step a.c. voltage up and down (11)
17 A quantity whose unit is the ohm (10)
18 Attracts a north pole (5, 4)
19 An electrical component with three terminals (10)

Down:

2 Used to deflect the electron beam in a cathode ray tube (8, 5)
3 A quantity whose unit is the volt (8, 10)
4 Used to switch on a large voltage with a small voltage (5)
5 Used to measure the potential difference across a component (9)
9 Used to vary the potential difference across a component (13)
10 Measures the current through a component (7)
13 A quantity whose unit is the ampere (7)
14 Transforms an input signal of 1 to an output of 0 (3, 4)

Examination style questions on Section 4

1. The diagram below shows a low-voltage lighting circuit.

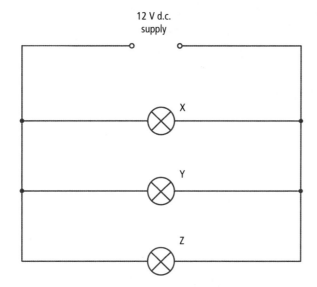

12 V d.c.
supply

a.　Indicate with a dot and the letter S, a point in the circuit where a switch could be placed that would turn off lamps Y and Z at the same time but would leave lamp X still lit.

b.　i)　Draw the circuit symbol for a component that would vary the brightness of lamp X.

　　ii)　Mark with a dot and the letter R where this component should be placed.

c.　Calculate the current in lamp Y.

d.　The current in lamp Z is 3.0 A. Calculate the resistance of this lamp.

e.　The lamp Y is removed.

　　i)　Why do lamps X and Z still work normally?

　　ii)　The current in lamp X is 1.0 A. Calculate the current supplied by the battery with lamp Y removed.

CIE 0625 November '06 Paper 3 Q8

2. Below is a sketch of some apparatus, found in a Science museum, which was once used to show how electrical energy can be converted into kinetic energy.

When the switch is closed the wheel starts to turn.

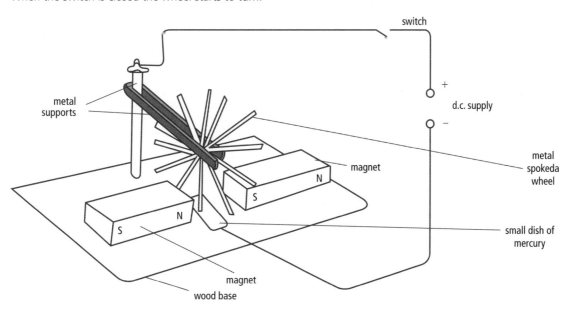

a. Explain why the wheel turns when the switch is closed.
b. Draw an arrow to show the direction of rotation of the wheel.
c. The d.c. motor is another way to convert electrical energy into kinetic energy. Draw a labelled diagram of a d.c. motor.
d. Describe how the split-ring commutator on an electric motor works.

CIE 0625 June '07 Paper 3 Q9

3. The diagram below shows an electrical circuit.

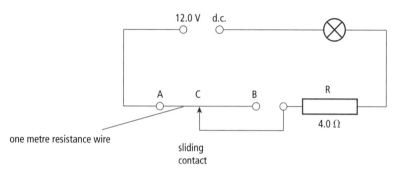

The resistance of the lamp is 4.0 Ω when it is at its normal brightness.

a. The lamp is rated at 6.0 V, 9.0 W.
   Calculate the current in the lamp when it is at its normal brightness.
b. The sliding contact C is moved to A. The lamp lights at its normal brightness.
   Calculate
   i)   the total circuit resistance,
   ii) the potential difference across the 4.0 Ω resistor R.
c. The sliding contact C is moved from A to B.
   i)   Describe any change that occurs in the brightness of the lamp.
   ii) Explain your answer to (i).
d. The 1 m wire between A and B, as shown in the diagram, has a resistance of 2.0 Ω. Calculate the resistance between A and B when
   i)   the 1 m length is replaced by a 2 m length of the same wire,
   ii) the 1 m length is replaced by a 1 m length of a wire of the same material but of only half the cross-sectional area.

CIE 0625 June '06 Paper 3 Q8

4. a. The diagram below shows an a.c. supply connected to a resistor and a diode.

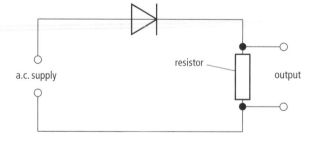

   i)   State the effect of fitting the diode in the circuit.

ii) Sketch graphs to show the variation of the a.c. supply voltage and the output voltage with time.

b. i) Draw the symbol for a NOT gate.
   ii) State the action of a NOT gate.

CIE 0625 June '05 Paper 3 Q9

5. a. The diagram below shows two groups of materials.

GROUP 1	GROUP 2
copper	plastics
iron	silk
gold	glass
aluminium	ebonite

i) Which group contains metals?
ii) Which group contains insulators?
iii) Write down the name of one of the eight materials above that may be charged by rubbing it with a suitable dry cloth.

b. Two charged metal balls are placed close to a positively-charged metal plate.

One is attracted to the plate and one is repelled.

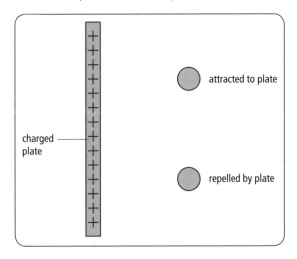

Write a + sign on the ball that is positively charged and a − sign on the one that is negatively charged.

c. State what is meant by an *electric field*.

CIE 0625 November '06 Paper 2 Q8

6. The diagram below shows a tube for producing cathode rays. The tube contains various parts.

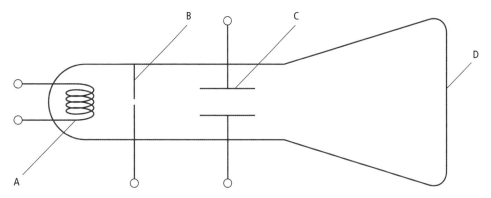

A spot is formed on the screen by the cathode rays.

a. What do cathode rays consist of?

b. Which part, A, B, C or D, must be heated to create the cathode rays?

c. i) Which part, A, B, C or D, is coated with fluorescent material?

   ii) What is the purpose of the fluorescent material?

d. A potential difference is applied between the two halves of part C.
   What effect does this have on the cathode rays?

e. Explain why there needs to be a vacuum inside the tube.

CIE 0625 November '05 Paper 2 Q11

7. a. The diagram below shows two resistors connected to a 6 V battery.

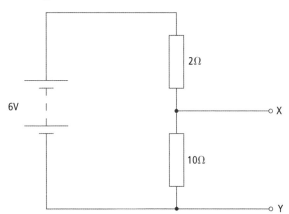

   i) What name do we use to describe this way of connecting resistors?

   ii) Calculate the combined resistance of the two resistors.

   iii) Calculate the current in the circuit.

   iv) Use your answer to a (iii) to calculate the potential difference across the 10 Ω resistor.

   v) State the potential difference between terminals X and Y.

b. The circuit below is similar to the circuit shown at the beginning of the question, but it uses a resistor AB with a sliding contact.

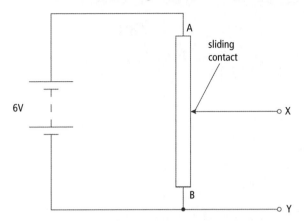

i)   State the potential difference between X and Y when the sliding contact is at
    1. end A of the resistor,
    2. end B of the resistor.

ii) The sliding contact of the resistor AB is moved so that the potential difference between X and Y is 5 V.

    On the second diagram, mark with the letter C the position of the sliding contact.

CIE 0625 June '07 Paper 2 Q9

5 Atomic physics

5.1 Radioactivity

KEY IDEAS

✓ There are three types of radioactive emission: α, β and γ, which have different characteristics

✓ The radiation all around us is called background radiation and it can be detected using a Geiger counter or photographic film

✓ In radioactive decay, the composition of the nucleus, and hence the atom, changes with the emission of α or β radiation

✓ The rate of decay of a radioisotope can be measured by its half-life

✓ Due to the dangerous nature of ionising radiation, sources of radiation must be stored in lead-lined containers and workers must be monitored for exposure

A radio isotope is an isotope of an element that has an unstable nucleus and can undergo **radioactive decay** by emitting **alpha** (α), **beta** (β) or **gamma** (γ) radiation (or a combination of the three). By emitting radiation, the nucleus becomes more stable. (For the definition of an isotope see section 5.2.)

Summary of the properties of alpha, beta and gamma radiation

	α	β	γ
What is it?	helium nucleus (2 protons and 2 neutrons)	fast moving electron	electromagnetic wave (*See section 3.2*)
Relative charge	+2	−1	zero
Relative mass	4	1/1800	zero

Atoms with unstable nuclei occur naturally and emit radiation, contributing to the **background radiation**. Higher doses of radiation can be dangerous because alpha, beta and gamma are all **ionising** and can cause cell damage. This means that when an alpha or beta particle or a gamma ray interacts with an atom, it can remove an electron from the atom. The atom becomes a positive **ion**. Alpha is the most massive and highly charged of the three types of radiation and is by far the most strongly ionising.

Summary of the behaviour of alpha, beta and gamma radiation

	α	β	γ
Ionising effect	strong	weak	very weak
Penetrating effect	Not very penetrating. Absorbed by a few cm of air or a few sheets of paper	More penetrating than alpha. Absorbed by a few mm of aluminium	Very penetrating. Not completely absorbed by lead and thick concrete.
Behaviour in electric and magnetic fields	Deflected by electric and magnetic fields	Deflected by electric and magnetic fields but in the opposite direction to alpha and by a greater amount	Not deflected by electric or magnetic fields

Effect of electric fields

Alpha particles are attracted to the negative terminal because they are positively charged. Beta particles are deflected in the opposite direction as they are negatively charged. Gamma rays are undeflected.

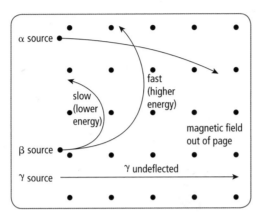

Effect of magnetic fields

▲ Effect of a magnetic field on alpha, beta and gamma radiation

Magnetic fields have different effects on alpha, beta and gamma radiation. The deflection of the alpha particles is given by Fleming's left hand rule (*see section* 4.5), with the direction of motion of the alpha particles as the current. They are deflected at right angles to the field and to their original direction. The beta particles are negatively charged and so they are deflected in the opposite direction to the alpha particles. They are deflected more because they have lower mass. Gamma rays are not charged so they are not deflected at all.

Examination style questions

1. Complete the following table about the particles in an atom

particle	mass	charge	location
proton	1 unit	+1 unit	In the nucleus
neutron			
electron			

    a.   Which of the particles in the table make up an alpha particle?
    b.   Which of the particles in the table is a beta particle?
    c.   What is the relative mass of an alpha particle?
    d.   What is the relative charge of an alpha particle?

CIE 0625 June '05 Paper 2 Q12

2. The diagram shows a beam of alpha, beta and gamma radiation. The beam passes between the poles of a very strong magnet.

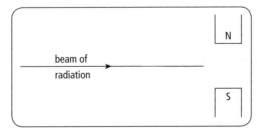

State the direction of deflection, if any, for each type of radiation.

Adapted from CIE 0625 June '06 Paper 3 Q11

3. a. Describe what happens when atoms are ionised by ionising radiation.

   b. Explain why alpha radiation is more ionising than beta radiation.

CIE 0625 November '06 Paper 3 Q11

Detection of ionising radiation

The Geiger-Muller (G-M) tube works by detecting the ions produced when alpha, beta or gamma radiation enters the tube. It is attached to a counter that registers a count each time a radioactive particle is detected.

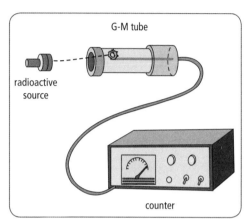

Photographic film is blackened by ionising radiation. The higher the number of radioactive particles incident on the film, the blacker it becomes (see film badge, page 156).

Random nature of radioactive decay

In a sample of radioactive nuclei, there is no way to predict which will be the next nucleus to undergo radioactive decay. Radioactive decay is a **random** process, although there is the same **constant probability** of decay for each nucleus. This is like throwing lots of dice at once. Each of the dice has the same probability of being a six (1/6) but you cannot predict which ones will be sixes each time you throw them.

Nuclear equations

When an unstable nucleus gives out radiation, its composition may change. This can be represented in a **decay equation**. On the left hand side of the equation is the original or **parent nucleus**, on the right hand side is the **daughter nucleus** and the radioactive particle that has been emitted.

Examples

(For an explanation of these symbols, see page 159.)

Alpha decay

$$^{222}_{88}\text{Ra} \longrightarrow {}^{4}_{2}\text{He} + {}^{218}_{86}\text{Rn}$$

Note that in alpha decay, the proton number **decreases** by 2 and the nucleon number **decreases** by 4.

Beta decay

$$^{14}_{6}\text{C} \longrightarrow {}^{14}_{7}\text{N} + {}^{0}_{-1}\text{e}$$

Note that the nucleon number is unchanged but the proton number **increases** by 1. This is because, in beta decay, a neutron is converted into a proton and an electron. The electron is then fired out of the nucleus but the proton remains. There is one less neutron and one more proton in the daughter nucleus.

Gamma decay

$$^{99}_{43}\text{Tc} \longrightarrow ^{99}_{43}\text{Tc} + ^{0}_{0}\gamma$$

In gamma decay, the number of neutrons and protons is **unchanged**. The gamma ray takes away some of the excess energy of the nucleus after it has emitted an alpha or beta particle.

Examination style questions

1. The nucleus of sodium-24 decays to magnesium-24 by the emission of one particle. In the equation below the symbol θ represents the emitted particle.

$$^{24}_{11}\text{Na} \longrightarrow ^{24}_{12}\text{Mg} + ^{x}_{y}\theta$$

a.   What is the value of x?
b.   What is the value of y?
c.   What is the emitted particle?

Adapted from CIE 0625 June '07 Paper 2 Q12c

2. The nucleus of uranium-238 decays to Thorium-234 by the emission of one particle. In the equation below θ represents the emitted particle.

$$^{238}_{92}\text{U} \longrightarrow ^{234}_{90}\text{Th} + ^{x}_{y}\theta$$

a.   What is the value of x?
b.   What is the value of y?
c.   What is the emitted particle?

3. The nucleus of iodine-131 decays to xenon-131 by the emission of one particle. In the equation below θ represents the emitted particle.

$$^{131}_{53}\text{I} \longrightarrow ^{131}_{54}\text{Xe} + ^{x}_{y}\theta$$

a.   What is the value of x?
b.   What is the value of y?
c.   What is the emitted particle?

Half-life

As a sample of radioactive material decays, its **activity** decreases with time. The activity is the number of radioactive particles emitted per second. As the number of parent nuclei decreases, the number of radioactive particles emitted per second decreases.

It is not possible to measure the time taken for a sample of radioactive material to completely decay because the activity never falls to zero. Instead, we measure the **half-life**.

The half-life of a radioactive isotope is the time taken for half the nuclei in the sample to decay, or the time taken for the activity of the sample to fall to half of its original value.

Example

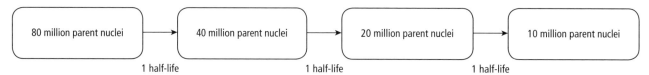

80 million parent nuclei → 40 million parent nuclei → 20 million parent nuclei → 10 million parent nuclei

1 half-life   1 half-life   1 half-life

Each time one half-life passes, the number of parent nuclei halves.

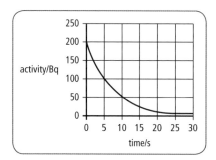

The graph shows how the activity of a sample of radioactive material varies with time. Note that it takes 5 seconds for the activity to halve from 200 to 100 counts per second and then a further 5 seconds to halve from 100 to 50 counts per second. How long does it take for the activity to fall from 50 to 25 counts per second?

Measuring half-life

The half-life of a radioactive isotope (radioisotope) can be measured experimentally, using a Geiger-Muller tube and counter to measure the activity. Before beginning the experiment, several readings of the background activity must be taken (the counts per second before the sample of radioactive isotope is placed in front of the detector). The **background count** must then be **subtracted** from all count rate readings. A graph of corrected count rate (count rate with background count subtracted) on the y-axis versus time on the x-axis is plotted and the half life found, as for the graph above.

Uses of radioisotopes

The use of a radioisotope depends on the type of radiation it emits and its half-life.

Medical tracers

A **gamma** emitting radioisotope with a **short half-life** (typically 6 hours) is injected into the patient and the gamma radiation that is emitted from his body is measured with a detector called a gamma camera. 'Hot spots' where gamma rays are being emitted at a higher rate show where there is a higher concentration of radioisotope. This allows the medical staff to diagnose a range of conditions, including cancer. Using a radioisotope with a short half-life ensures that the patient does not emit radiation for too long and gamma rays are easily detected outside of the body since they are very penetrating.

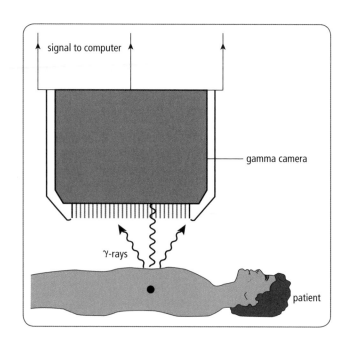

Carbon dating

All living things contain carbon. Many materials, such as cotton and wood, contain carbon because they came from once-living plants. Plants absorb carbon dioxide in the process of photosynthesis and convert it to carbohydrates, which the plant uses to make new plant tissue. However, not all of the carbon that the plant takes in is carbon-12 (the most abundant isotope of carbon). A certain percentage of carbon in the atmosphere is **carbon-14**, a radioactive isotope that emits beta radiation. The half-life of carbon-14 is **several thousand years**. When a tree is alive, it continues to take in carbon so the proportion of carbon-14 remains the same. However, once a tree has been cut down it stops taking in carbon from the atmosphere. The carbon-12 remains, but the carbon-14 decays and its activity drops over a very long period of time. Ancient artefacts containing animal or plant matter such as wood or cloth will have a lower proportion of carbon-14 than when they were made. The activity of the artefact can be measured and compared to what it would have been originally. For example, if the activity has halved, then one half-life has passed and an estimate of its age can be found.

Monitoring thickness

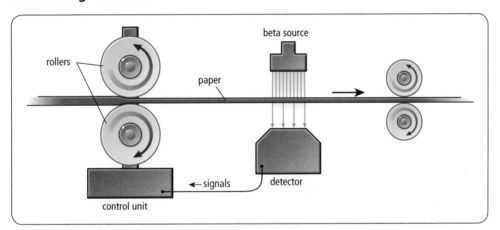

Paper mills use a beta-emitting radioisotope with a long half-life to ensure that the paper remains at the same thickness. The radioactive source is placed at one side of the paper and a detector at the other. Some of the beta radiation is absorbed by the paper but if the paper gets thicker more radiation is absorbed and the count rate at the detector decreases. The rollers adjust automatically to squash the paper and make it thinner again.

Detecting leaks

If there is a suspected leak in a water pipe, a gamma emitter of short half life (several hours) can be put into the water supply. A detector is held over the area of the suspected leak and the activity measured and compared to background activity. The gamma radiation would be partially absorbed by the metal pipe and only if there was a leak would the reading on the detector be significantly higher than background radiation. After several hours the levels of radiation would be safe again.

Medical therapy

Cobalt-60 is a radioisotope used frequently in radiotherapy. A beam of gamma radiation, emitted from a sample of cobalt-60 is focused with great precision onto a cancerous tumour in the body of the patient. The beam is rotated round the patient to reduce the dose to any one area of the body while the tumour receives a high enough dose to kill the cancer cells.

Safety precautions

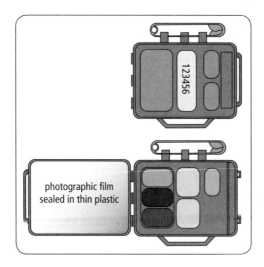

photographic film
sealed in thin plastic

▲ Film badge

Ionising radiation can **damage** or **kill cells** or, if changes take place in the DNA in the nucleus of the cell, cause mutations of genes and lead to cancer. Therefore, people who work with radiation must take safety precautions. Radiation workers wear **film badges**, which monitor the dose of radiation received, to ensure that it does not exceed safe levels. Radiotherapists handle radioactive sources in **lead** lined syringes and stand behind a lead screen.

Examination style questions

1. The graph below is a decay curve for an alpha-emitting radioisotope.

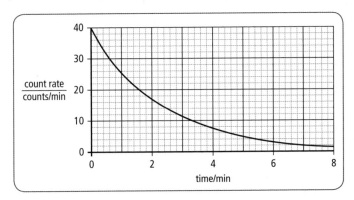

Showing your working on the graph, find the half-life for the radioisotope.

Adapted from CIE 0625 June '05 Paper 3 Q10a

2. A student uses a radioisotope that emits only alpha particles and has a long half-life to investigate the percentage of alpha particles that are absorbed by 3 cm of air.
   a.   Draw a labelled diagram of the experimental arrangement required to make the determination.
   b.   List the readings that the student should take.

Adapted from CIE 0625 June '05 Paper 3 Q10b

3. Suggest two uses for radioactive isotopes. For one of your examples, describe how it is used or draw a diagram to illustrate its use.

CIE 0625 November '06 Paper 3 Q11

5.2 The nuclear atom

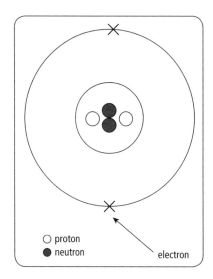

proton
neutron
electron

- The nucleus is in the centre of the atom and contains the **protons** and **neutrons**.
- Protons and neutrons are called **nucleons**.
- The electrons orbit the nucleus.
- The nucleus is **very** small compared to the size of the atom (about 1/10 000 of the diameter).

Relative masses and charges of the particles in an atom:

Name of particle	Relative mass	Relative charge
Proton	1	+1
Neutron	1	0
Electron	1/1800	−1

In a **neutral atom** the number of electrons orbiting the nucleus is equal to the number of protons in the nucleus. Therefore the positive and negative charges cancel out to give **zero** charge.

Number of protons in the nucleus = **proton number** = Z
Number of nucleons in the nucleus = number of protons and neutrons
      = **nucleon number**
      = **A**

A **nuclide** is an atom which is specified by its **proton number** and **nucleon number**.

We represent nuclides like this:

Example: Carbon-14

Number of protons = 6 = Z
Number of neutrons = 8 = A − Z
Symbol = C

$$^{14}_{6}\text{C}$$

Structure of an atom of carbon-14

This is a diagram representing carbon-14.

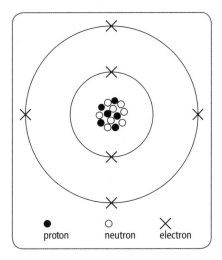

● proton   ○ neutron   ✕ electron

Isotopes

Isotopes of an element have the same number of protons in the nucleus of the atoms but different numbers of neutrons.

Example: carbon

An atom of the most common isotope of carbon, carbon-12, has 6 neutrons in its nucleus. An atom of carbon-14, shown above, has 8 neutrons in its nucleus. The atom is **neutral**, which means the overall charge on it is zero.

Carbon-12 has a stable nucleus, with the protons and neutrons held together by the strong nuclear force. Having two extra neutrons makes the nucleus of carbon-14 unstable and therefore **radioactive** (see section 5.1).

The discovery of the nucleus

The nuclear model of the atom was not widely accepted until the early part of the 20th century, when Rutherford and his research team gathered evidence for the existence of the nucleus by carrying out the following experiment.

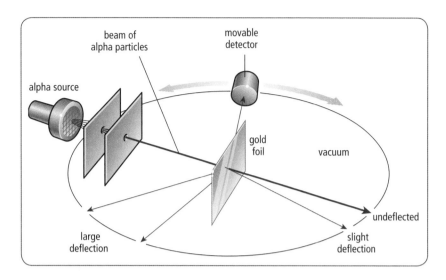

Small positive particles, called alpha particles (*see section 5.1*) were fired at a thin gold foil.

Diagram	Description	Conclusion
gold foil alpha particles	Most of the alpha particles went straight through the gold foil without changing direction.	This showed Rutherford that the atom was mainly space.
gold foil alpha particles	Some were deflected through small angles	This showed Rutherford that there was positive charge in the atom that repelled the positively charged alpha particles.
gold foil alpha particles	A few were deflected backwards, towards the alpha source.	The positive charge in the atom must be very concentrated to exert such a large force on the alpha particles. The mass and the positive charge are concentrated at the centre of the atom.

Examination style questions

1. Uranium-238 can be represented as $^{238}_{192}$U. How many protons, neutrons and electrons are there in an atom of uranium-238?

Adapted from CIE 0625 June '07 Paper 2 Q12

2. The diagram shows three alpha particles moving towards a thin gold foil.

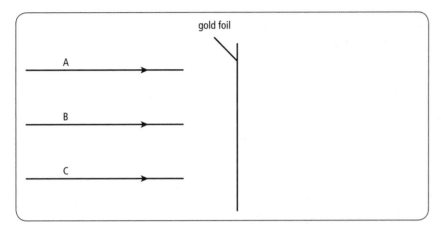

Particle A is moving along a line that does not come close to a gold nucleus.
Particle B is moving along a line that passes close to a gold nucleus.
Particle C is moving directly towards a gold nucleus.

a.   Complete the diagram to show the paths of particles A, B and C.

b.   State how the results of such an experiment, using large numbers of alpha particles, provides evidence for the existence of nuclei in gold atoms.

CIE 0625 June '07 Paper 3 Q11

3. You have learned about protons, neutrons and electrons

a.   How many of each are found in an alpha particle?

b.   How many of each are found in a beta particle?

Adapted from CIE 0625 June '07 Paper 2 Q12

Summary questions on unit 5

1. Fill in the blanks

The three types of nuclear radiation are _____, _____ and _____.

All are emitted from an unstable _____. The most penetrating type

of radiation is _____. The most ionising type of radiation is _____. Alpha

radiation cannot penetrate a sheet of _____. Beta radiation is absorbed by

a few _____ of aluminium. When entering an electric field, beta

particles will be _____. They are attracted to the _____ electric

pole. When entering a magnetic field alpha particles are _____ at a

_____ angle to their original direction and the direction of the field.

2. Match the statements

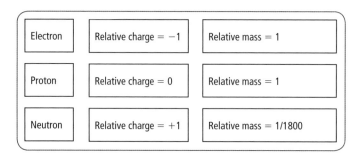

Electron	Relative charge = −1	Relative mass = 1
Proton	Relative charge = 0	Relative mass = 1
Neutron	Relative charge = +1	Relative mass = 1/1800

3. Draw a diagram to show what happened in Rutherford's α scattering experiment. Draw the gold nucleus and a line to indicate the path of the α particles before and after interaction with the gold nucleus.

Most α particles did not get close to a gold nucleus and did not experience a strong repulsive force.

Some α particles came close to a nucleus and were repelled by the positive charge.

Very few α particles hit a gold nucleus head on and were repelled strongly.

4. Complete the table

Isotope	Number of protons in nucleus	Number of neutrons in nucleus
$^{14}_{6}C$		
$^{3}_{2}He$		
$^{238}_{92}U$		
$^{231}_{95}Am$		